AF735988

CLASSICAL AND QUANTUM MECHANICS

# QUANTUM SPACETIME

## MIMICRY OF PATHS AND BLACK HOLES

# CLASSICAL AND QUANTUM MECHANICS

Additional books in this series can be found on Nova's website under the Series tab.

Additional E-books in this series can be found on Nova's website under the E-book tab.

CLASSICAL AND QUANTUM MECHANICS

# QUANTUM SPACETIME

## MIMICRY OF PATHS AND BLACK HOLES

MARCO SPAANS

nova publishers
*New York*

Copyright © 2015 by Nova Science Publishers, Inc.

**All rights reserved.** No part of this book may be reproduced, stored in a retrieval system or transmitted in any form or by any means: electronic, electrostatic, magnetic, tape, mechanical photocopying, recording or otherwise without the written permission of the Publisher.

We have partnered with Copyright Clearance Center to make it easy for you to obtain permissions to reuse content from this publication. Simply navigate to this publication's page on Nova's website and locate the "Get Permission" button below the title description. This button is linked directly to the title's permission page on copyright.com. Alternatively, you can visit copyright.com and search by title, ISBN, or ISSN.

For further questions about using the service on copyright.com, please contact:
Copyright Clearance Center
Phone: +1-(978) 750-8400 Fax: +1-(978) 750-4470 E-mail: info@copyright.com.

## NOTICE TO THE READER

The Publisher has taken reasonable care in the preparation of this book, but makes no expressed or implied warranty of any kind and assumes no responsibility for any errors or omissions. No liability is assumed for incidental or consequential damages in connection with or arising out of information contained in this book. The Publisher shall not be liable for any special, consequential, or exemplary damages resulting, in whole or in part, from the readers' use of, or reliance upon, this material. Any parts of this book based on government reports are so indicated and copyright is claimed for those parts to the extent applicable to compilations of such works.

Independent verification should be sought for any data, advice or recommendations contained in this book. In addition, no responsibility is assumed by the publisher for any injury and/or damage to persons or property arising from any methods, products, instructions, ideas or otherwise contained in this publication.

This publication is designed to provide accurate and authoritative information with regard to the subject matter covered herein. It is sold with the clear understanding that the Publisher is not engaged in rendering legal or any other professional services. If legal or any other expert assistance is required, the services of a competent person should be sought. FROM A DECLARATION OF PARTICIPANTS JOINTLY ADOPTED BY A COMMITTEE OF THE AMERICAN BAR ASSOCIATION AND A COMMITTEE OF PUBLISHERS.

Additional color graphics may be available in the e-book version of this book.

## Library of Congress Cataloging-in-Publication Data

Quantum spacetime : mimicry of paths and black holes / editor: Marco Spaans (Kapteyn Astronomical Institute, University of Groningen, Groningen, The Netherlands).
pages cm -- (Classical and quantum mechanics)
Includes index.
ISBN 978-1-63482-711-9 (hardcover)
1. Space and time. 2. Quantum theory. 3. Black holes (Astronomy) I. Spaans, Marco, 1969- editor. II. Series: Classical and quantum mechanics series.
QC173.59.S65Q85 2015
530.11--dc23

2015013326

*Published by Nova Science Publishers, Inc. † New York*

*Identity by mimicry*
*Alteration by observation*

*We are one, two and many*
*Our existence appears uncanny*
*Yet as we walk through time*
*And thus unfold this rhyme*
*Its words suggest to mean*
*What uncertainty may redeem*

# Contents

# Acknowledgments

The author is grateful to Elly van 't Hof and Pieter de Laat for many discussions on the nature of spacetime.

# About the Author

**Prof. Dr. Marco Spaans**
Kapteyn Astronomical Institute
University of Groningen
P. O. Box 800, 9700 AV Groningen
The Netherlands
Email: spaans@astro.rug.nl

Prof. Dr. Marco Spaans (PhD, Leiden University) is a full professor of theoretical astrophysics at the University of Groningen, The Netherlands. He has held positions at the Johns Hopkins University and Harvard University. In his spare time he likes to explore culinary secrets and fine wines.

# Preface

This monograph explores the foundations of quantum spacetime in a manner that is accessible to researchers and students as well as interested laymen that lack a strong background in quantum mechanics or spacetime studies, but are keen to learn. To serve this broad audience, various chapters offer introductions to the fields of quantum mechanics, topology and geometry, general relativity, black holes and wormholes, and cosmology; all without much advanced mathematics. As the reader progresses, conceptually challenging topics are built up. This monograph does not provide a complete overview of the various topics, but introduces those elements that are relevant for quantum spacetime. Toward the end, the text is more aimed at experienced researchers, but non-expert readers can find their way as well.

Different from other texts, this monograph derives the necessary background on quantum mechanics, topology and geometry, general relativity, black holes and wormholes, and cosmology from elementary and intuitive notions about spacetime. The central perspective of this book is that Nature expresses itself through spacetime and, to this effect, continuously (re-)constructs spacetime globally through the use of *paths*. In this, a path is what one expects it to be: a route from a location and time A to another location and time B. Spacetime is the union of time with three spatial dimensions, yielding four dimensions in total. Through four-dimensional spacetime, a path is then best viewed as an entire history of events. For example, one's life and all experiences in it constitutes such a path through spacetime, as a series of events.

An approach where paths through spacetime carry physical phenomena is somewhat unconventional. This is intended to create the right mindset for the reader when it comes to the mysteries of quantum mechanics, whatever his/her

# Chapter 1

# A Few Reflections on the Nature of Spacetime

Space and time are as fundamental physically as they are pertinent to daily experience. Everyone has an intuitive association with space and time. Space measures location in some volume. Time expresses change. This realization may often be implicit, but knowing where one is and what time it is are deeply embedded in our consciousness. This is also true for any particular object that one is interested in, like a set of keys, a favorite pen or a grocery list. Such items require a placement in space and time if they are to exist with respect to each other.

A priori one may consider space and time to be separate notions. However, as every traveler can appreciate, a journey from A to B appears to take one from some moment in time to another while traversing any distance. Conversely, even staying in a single location requires the lapse of time to assess that one has not actually moved. As such, it is natural to not impose a rigid distinction on space and time. Rather, one can view distance and duration as the two sides of a single coin, jointly describing the history of material objects.

Indeed, since the ideas of Einstein, Minkowski and Lorentz (to mention only a few) about a century ago, it has become clear that space and time should be perceived as part of a more fundamental entity called spacetime. This four-dimensional structure, also referred to as four-space by the more mathematically inclined, constitutes the arena in which physical phenomena occur. The name four-space is actually a good one, because it highlights the fact that space and time are not to be seen separately.

But there is more. Spacetime itself is not static and absolute, like it was perceived to be in the time of Newton. Rather, it is a dynamical continuous fabric without an absolute measure of length or duration. Chapters 3 and 4 discuss, through Einstein's special and general theory of relativity, how absolute properties of space and time are untenable in the light of empirical evidence and purely logical arguments. For now, one should appreciate the simple fact that spacetime is a dynamical continuum. In other words, spacetime is not a fixed stage on which material objects perform physical phenomena. On the contrary, it is a four-dimensional structure that changes its shape and connectivity, thereby affecting its constituent tangible parts. It does so for two reasons.

First of all, the speed of light in vacuum is constant whatever the motion of an observer. Because a speed has the physical dimensions of distance divided by duration, space and time coordinates must somehow conspire to render the speed of light invariant. Consequently, and without having to be mathematically explicit, measurements with rulers and clocks must lose any absolute meaning.

Second of all, any mass gravitates and this property expresses itself through how spacetime curves in the presence of a massive body, like the Sun or the Earth. This stretching and bending of spacetime, as if it were made from rubber, again indicates that any fixed measure of distance and duration is not applicable to our universe. Chapters 3 and 4 discuss some of the special (physical and mathematical) properties that spacetime possesses when it is a dynamical entity. This while the existence of exotic objects like black holes and wormholes are treated in chapter 5, and the very origin and expansion of our cosmos is highlighted in chapter 6. All these notions derive from a four-space continuum that provides distinct paths for material objects to follow. The idea that we live in a dynamical spacetime, beautifully formulated by Einstein in 1915, has now reached the respectable age of about 100 years. This while many experimental and observational studies have confirmed unambiguously that space and time form one fluid four-dimensional fabric.

Nevertheless, chapter 2 forms the start of this monograph, and focuses on an equally fundamental but even more mysterious property of our universe: its quantum nature. At this stage one merely needs to realize that Nature is fundamentally *undetermined*. That is, the outcome of interactions, so the occurrence of events, is not unambiguous. Rather, somewhat like everyday life, things are uncertain. A simple example is the decay of radio-active material, like nuclear waste. Such decay occurs on a characteristic time scale, called the half-life. After one half-life, about half of the nuclear material has decayed to a different, and usually less harmful, substance. However, the decay of an individual radio-active atom, so when perceived as a single event, is not an exactly determined notion. That is, the decay of a single atom may take significantly longer or shorter than the characteristic half-life. This while a large number of atoms, like a container of nuclear waste, has a very well determined statistical average.

Clearly, the history of an individual radio-active atom and the average of an entire collection are not necessarily the same. This despite the fact that all such decaying atoms are intrinsically equivalent, subject to the same rules for decay. Apparently, different routes through spacetime are possible for otherwise identical objects that are governed by a single set of natural laws. These spacetime routes are to be perceived as entire four-space histories. Indeed, Nature knows precisely how things happen, but not exactly what the results are...

In fact, any particle (like an electron or proton) is fundamentally undetermined in space and time. This uncertainty expresses itself pertinently, and in an easy to imagine fashion, through the inability of an observer to determine exactly and simultaneously where a particle resides and with what speed it travels. To be more precise, there is a reciprocal relation between the measurement uncertainty in location and the one in speed (Heisenberg's uncertainty principle, discussed in the next chapter).

To this effect, imagine a photographer trying to take a picture of a moving object. His/her first goal is likely to make a sharp image, using a very short shutter time. The location of the object is then very well determined. However, any indication of the object's speed is difficult to assess. Conversely, a longer shutter time yields a blurry image, with a poorly known location. Still, the extent of the now fuzzy object and the shutter time can be used to obtain an estimate of the object's speed.

Consequently, no particle can be localized precisely in spacetime because any velocity[1] measurement requires information on the distance traveled by a particle during a time interval. i.e., a particle's speed is the rate of change of its position. Obviously, trying to make the particle location exact just leads to a very uncertain speed and one therefore has no idea where the particle ends up

[1]This is a vector quantity, so it has a magnitude and a direction.

later in time. Vice versa, any well determined speed requires significant patches of length and duration and thus an undetermined stretch of a particle's trajectory through spacetime.

The observed particle in question thus appears to have different options for its movements. These depend on the specific actions of the photographer, who is also an observer. This sounds very strange and is precisely why quantum physics, usually referred to as quantum mechanics (as it is in this monograph), is so mysterious. The measurement uncertainty that pertains to an observed object, a fact that is very plausible by itself, is also tied to the behavior of the actual observer. That is, there is an interaction between the two that amounts to the concept that any form of observation invariably leads to a change in what is observed. Indeed, to observe is to affect! In this, an observation constitutes any form of comparison, i.e., any action that allows one to speak of both an observer and an observed. When viewed in terms of paths, this means that there are many different routes or histories that a particle can follow through spacetime, in response to the actions of an observer. Hence, the behavior of an observed entity in relation to an observer equates to there being a multitude of paths for any *joint* observer-observed system to experience as an expression of spacetime.

These arguments imply that to know how the spacetime continuum is built up in great detail, so in a point by point or actually a path by path manner, might be a challenge. Still, all four-space paths together do build up spacetime and thus automatically represent the quantum interactions between all material observers and observed in the universe. Hence, when four-space paths are perceived as intrinsically dynamical entities then they can facilitate the very quantum uncertainty effects that are experienced by material objects.

Evidently, such a marriage between spacetime and quantum mechanics is then highly desirable if one wishes to understand how the very concepts of space, time and matter come about at all. Unavoidably, the spacetime continuum takes central stage and is endowed with quantum properties. Four-space subsequently transfers its quantum properties to particles, rendering the quantum nature of matter a spacetime consequence. In this, the dynamical four-space continuum possesses a notion of uncertainty because it is built from individual paths that are themselves not precisely localized in space and time.

The latter is for the simple reason that one may suspect immediately. Any path goes from some spacetime point A to some spacetime point B by its very nature, while the spatial and/or temporal distance between A and B can be both small and large, depending on an observer's circumstances. Furthermore, there is more than one route that leads to Rome, so to speak. Particles can follow a multitude of paths, so enjoy different histories, between A and B if such is the structure of quantum four-space.

This approach constitutes a so-called path quantization of spacetime. The term quantization indicates that there is a discrete building block, the path in this case, from which spacetime is constructed. Such a perspective leads in part to the title of this monograph and the path is the central object in this work. Intuitively, a path driven philosophy relates strongly to human experience. There are many paths that one may follow in life, many four-space histories to experience. The question is then: how do choices and actions create and affect the paths that collectively form the spacetime that one occupies?

The basic constituents of our universe are therefore paths, in a pertinent physical and mathematical sense, as fibers in a fabric. Consequently, one can consider all possible paths that can express the structure of spacetime, also those that are clearly distinct from everyday experience and that make full use of four-space. e.g., a path that passes into a black hole or through a wormhole, if it is stable. One can even think of a path that loops through time to connect two regions of four-space. As if there is a hidden door connecting two different histories (so realities).

This sounds exotic, and it is, but such paths are shown to have deep physical meaning in the following chapters. Paths of all sorts allow two four-space points A and B to be connected in distinct physical manners. Through such paths spacetime becomes *multiply connected* because there are many different routes that connect any point A to any point B through any of the four dimensions. It is therefore crucial to realize, challenging as this may be, that in four dimensions paths constitute entire histories of objects and that many such paths both co-exist and interact with each other to build up the fabric of four-space. The rest of this monograph explores that idea and constructs a theory based upon it. Because paths are non-local objects in space and time, a lot is asked from the abstraction ability of the reader in four dimensions. The global spacetime that results, of a multiply connected nature and with many realities for matter to enjoy, hopefully rewards such efforts.

# Chapter 2

# Fundamentals of Quantum Mechanics

## 1. The Double Slit Experiment

The best way to grasp the mystery of quantum mechanics is through the double slit experiment. So here follows its description. Imagine two thin slits, very close together, that are typically used to generate interference patterns in light waves. For a light source behind the double slit and a screen in front of it, it is apparent that light waves travel slightly different path lengths through the two slits from the source to various points on the screen. Waves are not localized in space and time. This therefore leads to an interference pattern through the Huygens principle. A wave that reaches either slit becomes a source of a spherical out-going wave, and the superposition of such waves leads to a a distinctive pattern of peaks (constructive interference) and valleys (destructive interference).[1]

If one does this experiment with electrons, then no interference pattern should result if the electrons behave as particles. After all, particles are localized according to Newtonian mechanics and should move ballistically through either of the two slits. This yields two well defined pile-ups behind either slit depending on which slit the particles pass through by chance, as the source shoots

[1]Of course, it matters what the slit width and the wavelength of light are. For the sake of argument, one may simply take these to be equal.

of electrons randomly in the general direction of the two slits. This is unlike emitted waves that move through both slits because of their extended nature.

However, as many experiments have shown, electrons create an interference pattern as well. That is, quantum particles can also move through both slits as one single entity. The unavoidable conclusion is then that quantum particles (electrons, protons, etc.) behave like regular particles *and* waves. This is called particle-wave duality. The core of the quantum mystery is therefore that quantum particles, as the quantization of matter, possess wave properties in the sense that they can move simultaneously along different paths between any spacetime points A and B.

In the case of the double slit experiment, there are many paths for a particle to pass through one slit and many to pass through the other. This while both of these collections of paths are traveled along by any quantum particle. These two multitudes of paths, signifying the quantum phenomenology of the double slit experiment, then cause interference effects between any quantum particle and itself, as well as between different quantum particles.

The essential aspect here is that there are different spacetime paths associated with either slit and that this multiply connected nature of four-space imposes itself on quantum particles by forcing the latter to follow all available routes. In this, the two slits are material entities that merely bring out the underlying path structure of spacetime by the tangible obstruction that the material around the slits constitutes for traveling particles. The very existence of multiple spacetime paths between any points A and B is the truly fundamental and intrinsic property of four-space. Hence, that notion is explored in depth in the remainder of this monograph.

## 2. Schrödinger's Cat and Superposition: the Entangled Observer and Observed

Another manifestation of quantum mechanics is called Schrödinger's cat. Imagine a box with a cat and a vial containing poison. The vial can be broken, and the poison released, by a quantum process with an uncertain outcome. One can think again of the decay of a radio-active atom, as mentioned above, with a certain half-life. The decay can trigger a system that breaks the vial with poison. It is uncertain when the atom decays exactly. Hence, the release of the poison is not precisely known either.

This quantum uncertainty dictates that the cat is in a so-called *superposition* of two possible states: alive and dead. This superposition seems somewhat odd, but it can be understood in terms of paths. Because of the uncertainty involved in the cat's surroundings, dictated by radio-active decay, there are different possible histories. Some of these lead to death and others to the continuation of the cat's (mice chasing) existence.

Just as the electrons in the double slit experiment follow different paths, so can the occupants of the box undergo various changes with the passing of time. What these events are, remains unobserved as long as the box is closed. However, it is undeniable that the cat, the poison vial and their interaction through nuclear quantum processes, follow multiple paths through spacetime all the same. Only when the observer opens the box, and interacts with these observable entities, is one of the possible histories chosen. Observer and observed are said to be *entangled*, i.e., they are joined in four-space through the box, and the cat's observable fates are superposed.

The entanglement of states therefore comes about because there are two complete sets of events, two ensembles of four-space paths labeled "dead" and "alive", of which one can be eventually selected (so observed) if the observer chooses to do so. Both histories have meaning, and distinct yet equivalent merit of their own, but one history cannot exist without the other. This entanglement through different four-space paths is a crucial property of our universe.

Furthermore, superposition and the interaction between the observer and the observed are unambiguously demonstrated by a two-electron system, as follows. Electrons are spin $\frac{1}{2}$ particles. This means that they have the wonderful property that one gets the particle back in its original orientation only after two full rotations by 360 degrees, rather than one for spin 1 particles. The (rather tantalizing) nature of spin $\frac{1}{2}$ is returned to further below. For now, one should merely realize that through superposition one can make a system with two electrons in which one has the spin orientation $+\frac{1}{2}$ and the other $-\frac{1}{2}$. The reason for opposite orientations derives from Pauli's exclusion principle. It dictates that two identical spin $\frac{1}{2}$ particles are not allowed to enjoy the same quantum state simultaneously. Furthermore, because electrons are identical particles one cannot know which one is which in the spin superposition. Consequently, both electrons are in the $+\frac{1}{2}$ and $-\frac{1}{2}$ spin state with equal probability until an observer attempts an actual measurement and forces the issue.

To highlight the importance of paths hiding in this thought experiment, imagine that one moves one of the two electrons far away, from the set-up point A to some point B. After separation, say by as much as 10 light years, one can perform the measurement. Suppose that the remaining electron in point A is found to be in the $+\frac{1}{2}$ state. It must then follow that the other electron in the far-away point B is instantaneously in state $-\frac{1}{2}$, because of quantum entanglement. Similarly, if the A electron is detected to be in the $-\frac{1}{2}$ state, then an observer in B finds his/her electron to be in the $+\frac{1}{2}$ state upon subsequent measurement.

Einstein abhorred such spooky action at a distance. However, it is the reality of our universe and many experiments have confirmed this entanglement, for distances more modest than 10 light years of course. Rather than information traveling faster than light, which would violate causality, it is once more the role of paths that causes the effect, in the following manner.

As one moves the electrons apart, their very superposition requires there to be two distinct paths between A and B. That is, one has two properly causal histories through spacetime for the two separated superposition states (A in $+\frac{1}{2}$, B in $-\frac{1}{2}$) and (A in $-\frac{1}{2}$, B in $+\frac{1}{2}$). The action of an observer in point A or B then simply chooses one of these two global chains of events through his/her interaction with either observed electron. Any subsequent observation on the electron in point B or A merely expresses the same *global* information. Again, one is confronted with a sense of path multiplicity, as the expression of how non-local processes shape our universe when it comes to quantum measurement. Hence, it is prudent to explore the fundamental uncertainty in measurements that is so typical of our cosmos.

## 3. Quantum Measurement Uncertainty: Paths

If one wants to know by what route an object travels, then one needs to be able to measure. Therefore, imagine that one has a measuring stick in this thought experiment, like a ruler. The ruler has a limited accuracy. This measuring uncertainty can be small, but never zero. This makes good sense because, as everyday experience shows, there is always some uncertainty in measurements, from the GPS locations of cars to the placements of bricks in a wall. Before starting the thought experiment, it is important to realize that the measuring uncertainty, one can call it $\delta l$ in units of cm, determines how precisely one can write the tick marks on the ruler in the first place. So, if $\delta l$ is 1 cm then one can measure the length of a favorite pair of trousers to no better than about $\pm 1$ cm.

Imagine now that one has a ball with a GPS transmitter rolling down a soccer field. With a radio detector to make images and the ruler one can then determine, more or less, where the ball (or actually the transmitter) is. If one makes the ball roll across the field many times, each time placing it at the same position, as well as possible, and aiming it at the same target, as accurately as possible, then one will not measure *exactly* the same path for the ball. This is obvious because the limited accuracy of the ruler will prevent one a) from placing the ball exactly at the same starting point every time, b) from hitting the exact same end point with it, and c) from measuring the same path in between.

To push matters further, imagine doing this experiment with smaller particles, like the oxygen molecules one breaths, using a microscope, or even fast protons, like in the particle accelerator at CERN in Switzerland. If an observer does so with great dedication, and technological skill, then he/she will have to reach the same conclusion. Measured trajectories in repeated experiments with traveling particles do not match exactly, even though the routes can be very similar, if the ruler has an uncertainty $\delta l$ that is larger than zero. Hence, one can deduce, in general, that between any two spacetime points A and B an object appears to travel along many different paths as far as an observer with a ruler is concerned.

It is important to realize here, both for this thought experiment as well as in reality, that the points A and B can be at any distance from each other. Indeed, one can think of a route as being made up from many shorter paths that take one from four-space point A to four-space point B along many stops or as one long haul that goes from A to B. Either way, A and B can be as faraway or nearby in spacetime as one wants. Similarly, there is nothing special about the points A and B either. So the conclusions reached pertain to any spacetime location that can serve as an origin and/or destination.

This is a good moment to think about how small one can make $\delta l$ for our universe. There are three fundamental constants in Nature that are helpful here. There is Planck's constant $h$ from quantum mechanics, the speed of light $c$ from Einstein's special relativity theory, and Newton's gravitational constant $G$ from Einstein's general relativity theory. It is straightforward to see that these three constant numbers can be combined into a quantity that has the units of length in one and only one manner. This is called the Planck length $l_P = (\frac{hG}{2\pi c^3})^{1/2}$. The $2\pi$ in this formula indicates the quantum property that matter can behave like particles as well as waves (recall the double slit experiment above). Clearly, waves are basically sine and cosine functions with periods of 360 degrees. Two further unique measures are the Planck time $t_P = l_P/c$ and the Planck mass $m_P = (\frac{hc}{2\pi G})^{1/2}$. The constants $h$, $c$ and $G$ have been measured in the laboratory, and $l_P \approx 1.6 \times 10^{-33}$ cm, $t_P \approx 5 \times 10^{-44}$ s, and $m_P \approx 2 \times 10^{-5}$ g.

The Planck length is tiny, many times smaller than an atomic nucleus. However, it is *not* zero and constitutes the certainty with which Nature can measure. At such a level of uncertainty, although precise enough from trousers all the way down to the quarks inside the nuclei of atoms, one thus reaches a fundamental barrier. Even Nature itself, which can only work with its own universal constants, does not know the precise paths taken by objects to better than $\pm l_P$.[2]

It is good to let this sink in... Uncertainty about how one goes from A to B is an intrinsic and fundamental property of our universe. The fact that $l_P$, $t_P$ and $m_P$ are not zero renders any path, as a fiber of spacetime for matter to travel along, *quantum* in nature as well. That is, spacetime histories are themselves not better defined than by the discrete units $l_P$, $t_P$ and $m_P$. This in turn implies that the fabric of spacetime itself is uncertain because no individual path in it can be described exactly in terms of distance, duration and the mass that it guides. Consequently, quantum spacetime is built from distinct histories that are intrinsically uncertain yet should be collectively meaningful as the carriers of reality.

Before exploring such a fundamental realization further, and to construct this quantum spacetime of paths, it is good to focus first on particles traveling through spacetime. This allows one to learn of a powerful (somewhat mathematical) quantum mechanical framework that serves the discussions in subsequent chapters. The step to quantum aspects of four-space itself then follows naturally.

[2]This argument even holds if $h$, $c$ and $G$ vary with time and/or location, i.e., if they are not truly fundamental constants. As long as the resulting values of $l_P$, $t_P$ and $m_P$ are non-zero, quantum uncertainty is unavoidable.

# 4. Many Distinct Yet Equivalent Paths to Express Quantum Effects

A way in which one can express the path aspect of quantum mechanics locally, for individual particles, is with operators. An operator acts on whatever it is followed by. So $\mathbf{p} = \frac{h}{2\pi i}\frac{d}{dx} = -i\frac{h}{2\pi}\frac{d}{dx}$ is proportional to the derivative of a function $F(x)$ when writing $\mathbf{p}F = \frac{h}{2\pi i}\frac{dF}{dx}$. The imaginary number $i = (-1)^{1/2} = -1/i$ is explained further below and boldface is used to indicate operators. In this, $x$ denotes location along the x-axis for a simple one-dimensional space. Similarly, one may define the operator $\mathbf{x} = x$, which yields $\mathbf{x}F = xF$. The operator $\mathbf{p}$ is called the momentum operator and the operator $\mathbf{x}$ is referred to as the location operator. The momentum $p$ is of course equal to the mass of an object times its velocity. Newtonian mechanics teaches one that momentum rather than speed is what determines particle dynamics, and thus motivates the quantum operator $\mathbf{p}$.

Because particles behave as waves, and also have a sense of location and momentum, one can consider the plane wave $ae^{2\pi i(px-Et)/h}$ for a quantum particle with amplitude $a$ moving freely without any interaction. This introduces a phase factor $f = e^{2\pi i(px-Et)/h}$ of the wave as a function of time $t$, energy $E$, location $x$ and momentum $p$ when traveling along the x-axis of some arbitrary coordinate system. As desired, any spatial and temporal derivatives of $f$ are proportional to $f$ and thus preserve the overall phase structure of quantum matter. With the concept of a phase factor $f$ one then has the so-called wave function $\psi = af$. This is a very special function because it is the mathematical embodiment of particle-wave duality, as follows.

The phase factor is a complex number and expresses the wave-like property of particles because any complex number $z = re^{ig}$ can be written as $z = d + ie = r[cos(g) + isin(g)]$. i.e., $z$ is a combination of sinus and cosine waves, where $i = e^{i\pi/2}$ amounts to a phase of 90 degrees in the so-called complex plane $d - e$, with $d$ along the horizontal and $e$ along the vertical direction. The pre-factor $r = (d^2 + e^2)^{1/2}$ follows from Pythagoras. One adds two complex numbers $z_1$ and $z_2$ through $z_1 + z_2 = (d_1 + d_2) + i(e_1 + e_2)$. Indeed then, the imaginary axis $ie$ allows a particle to follow paths $x(t)$ that modify its phase $2\pi i(px - Et)/h = g$ through a series $z_1(g_1), z_2(g_2), \ldots$ and superpose it into a wave *package* $z_1(g_1) + z_2(g_2) + \ldots$ that acts as a particle.

Note that complex numbers are not well ordered. For example, $2i > i$ seems correct, but division by $i$ yields $2 > 1$ and multiplication by $i$ given the inconsistency $-2 > -1$. Hence, only real numbers that derive from complex ones constitute well defined observables in some four-space history. To know with what probability a quantum particle resides between A and B along the $x$-axis, one first considers the real square $\psi\psi^*$. Here, $\psi^*$ is the complex conjugate of $\psi$, i.e., through $z^* = d - ie = re^{-ig}$ (so conjugation replaces $i$ with $-i$ everywhere). Then one computes $\int_A^B \psi\psi^* \; dx$. The so-called probability amplitude $\psi\psi^*$ should be properly normalized to unity, i.e., $\int_{-\infty}^{\infty} \psi\psi^* \; dx = 1$ if the particle can travel along the entire $x$-axis.

The use of the so-called squared modulus $\psi\psi^*$ as a probability density, called the Born rule, yields a real number befitting an observable quantity. In general, superposition of quantum particles $j$ can now proceed naturally through $\Psi = \psi_1 + \psi_2 + ...$ for individual wave functions (packages) $\psi_j$, like the two-electron system above. The resulting overall probability amplitude is then $\Psi\Psi^*$. But there is more to this. The need for the square of $\psi$ or $\Psi$ has a deeper meaning in terms of multiple paths, as follows.

If one wants to know an observable property of a particle, then it is the interaction between the observer and the observed, which are both quantum objects in nature, that yields this. Hence, for any observer and any quantum particle, one requires a path (an entire history) for each one to allow the four-space propagation of the wave function associated with their mutual interaction, like for Schrödinger's cat above. These two four-space histories, actually two sets of paths that superpose into an observer and an observed, have to be *distinct* because the observer and the observed both need to have an individual identity through their spacetime histories. This while the two sets of four-space paths also need to be *equivalent* because there is no fundamental difference between the particle affecting the observer or the observer influencing the particle. i.e., an observer can always adopt the role of the observed, and vice versa, whatever consequences this has for spacetime dynamics!!

This is a crucial perspective to embrace. The very existence of observables associated with the interactions between an observer and an observed requires multiple four-space paths that are *distinct yet equivalent*. It may help to repeat this idea a few times for one-self. It is the heart of quantum spacetime and the very construction of the latter comprises the identification of such distinct yet equivalent four-space paths (i.e., entire histories). As a first step, it is good to apply this idea to the existence of quantum observables.

Having just two distinct yet equivalent paths, so one history for an observer and one for an observed, already assures that the real self-interaction $\psi^2 = \psi\psi^*$ defines an observable existence for the particle in relation to its observer. Conversely, the wave function $\psi$ alone, even though a superposition along many paths, is without any intrinsic identity. It is only through *comparison* with a distinct yet equivalent set of paths associated with an observer, being $\psi^*$, that an actual observation can result. Indeed, one has that $\psi \neq \psi^*$. While $\psi$ and $\psi^*$ can also be obtained from each other, and thus represent equivalent information, by a complex conjugation. Of course, the mathematical operation of conjugation merely flips the sign of the phase in $f$, as if an observer and an observed jointly look at each other's reflection.

Indeed, the fundamental underlying principle is that observers and observed can adopt each other's role in four-space, while both are necessary as individual entities to define reality. That is, the act of observation singles out a particular history, as in the examples of Schrödinger's cat or the two separated electrons in superposition, through the interaction between observer and observed. This requires both the observer and the observed to individually carry the intrinsically joint event that results from their mutual interaction. The latter dual notion of existence can only occur when observer and observed are described by four-space paths that are both distinct and equivalent.

This is challenging to imagine. Still, the idea that any interaction between observer and observed requires the comparison of distinct yet equivalent paths through four-space to carry quantum entanglement is very important for the reader to remember. This perspective is made more precise as a guiding physical principle in subsequent chapters. For now, it is best to appreciate the fact that such paths between any spacetime points A and B need to (and can) exist, even though distinct yet equivalent may sound like a paradoxical phrase.

To lift a tip of the veil, the imaginative reader may already realize that equivalent is not the same as identical. Furthermore, any comparison between histories can be made locally, so through microscopically short stretches of paths, and globally, so by considering paths between spacetime points A and B that are macroscopically separated from each other. Therefore, two paths may be the same locally for selected events. This while they are not the same globally, rendering them distinct in four-space when all histories leading to those selected events are considered. This way, one can have his/her cake and eat it too when it comes to observer-observed entanglement in quantum spacetime. Of course, one actually needs to *construct* such a universe of many distinct yet equivalent paths that allows observers and observed to always adopt each other's role.

## 4.1. To Commute or Not to Commute

First however, there is more to do with operators. Obviously, $\mathbf{p}f = pf$ holds and the momentum operator returns the momentum of the quantum particle. The same pertains to the particle's location through $\mathbf{x}f = xf$. Interestingly, the commutator defined as $[\mathbf{p}, \mathbf{x}]f = \mathbf{p}\mathbf{x}f - \mathbf{x}\mathbf{p}f$ does not equate to zero. This is because the order of determining the location and then the momentum of a quantum particle yields a different result than when one first determines its momentum and subsequently its location. Straightforward differentiation of the $f$ and $xf$ terms in $[\mathbf{p}, \mathbf{x}]f$ proves this. i.e., the location and momentum operators do not commute. Similar conclusions are reached when one works with time $t$ and the energy operator $\mathbf{E} = i\frac{h}{2\pi}\frac{d}{dt}$. In fact, for particle mass $m$, energy and momentum are related by $E = p^2/(2m)$. Therefore, $i\frac{h}{2\pi}\frac{d}{dt} = -(\frac{h}{2\pi})^2/(2m)\frac{d^2}{dx^2}$ is the operator expression of how the phase factor $f$ generates particle properties.

The reason for the non-commutativity result is intuitive to understand. As one a) determines the momentum accurately with the operator $\mathbf{p}$, the location of the quantum particle becomes undetermined. Conversely, if one b) first determines the location accurately, then the subsequent determination of the momentum is unsure. In effect, the two operations a) and b) consider two different paths. As one performs these observations repeatedly during a time interval, then on each occasion part of the particle trajectory is either extended erratically and subsequently specified or one fixes a short stretch of the particle path and then randomly changes it. Clearly, the resulting two paths cannot be exactly the same at any given time.

In any case, the order in which one performs location and momentum operations, so the way in which stretches of paths are chosen, matters. The difference between such operations is exactly $\frac{h}{2\pi i}$, so scales with the natural unit of Planck's constant (angular momentum in g cm$^2$ s$^{-1}$). The smallest difference among all paths comprising quantum spacetime is therefore one quantum $\frac{h}{2\pi}$. This is the beauty of quantization in terms of paths when it comes to the combination of a particle's location $x$ in cm (or time $t$ in s) and momentum $p$ in g cm $\mathrm{s}^{-1}$ (or energy $E$ in erg), jointly present in the phase factor $f$ and normalized to $\frac{h}{2\pi}$.

The location and momentum operators thus trace out a particle's history together. The task at hand now is to capture the underlying path structure of quantum spacetime as an expression of *all* interactions among observers and observed. In such a world, a single quantum has the deeper meaning of a discrete change in the four-space fabric itself. So distinct yet equivalent paths are not only the fundamental and dynamical entities that drive quantum reality, they are also countable in the Planck scale units $l_P$, $t_P$ and $m_P$.

## 4.2. Heisenberg's Uncertainty Principle

Another step toward the goal of quantum spacetime lies in Heisenberg's uncertainty principle. Imagine a particle moving along a single direction, with its location indicated by $x$. Heisenberg's principle then relates the degree of uncertainty in the object's location, $\Delta x$, to the uncertainty in its momentum along the same direction, $\Delta p$, as $\Delta x \Delta p \geq \frac{h}{2\pi}$. One can be quite precise about what the uncertainty $\Delta$ means. Often the standard deviation is used, from the mathematical field of statistics, which leads to another factor of $\frac{1}{2}$ on the right hand side above and directly relates to the wave-like properties of matter.

However, for the present discussion it only matters that there is a measure of uncertainty, one intrinsic to Nature itself. Indeed, Planck's constant has units of length times momentum, or equivalently energy multiplied by time. It is then only the product of $\Delta x$ and $\Delta p$ that has the same units as Planck's constant $h$ and that embodies the non-commutativity property of the operators $\mathbf{p}$ and $\mathbf{x}$ discovered above. One thus has a relation where one measurement uncertainty is inversely proportional to the other. This is a *local* approximation to the changing nature of the paths connecting any two arbitrary spacetime points A and B. Hence, one still needs to find a *global* formulation of changing paths that act as dynamical entities under the entanglement of observers and observed. At the least, one needs to go from one-dimensional displacements $x(t)$ to entire four-space paths.

## 4.3. Feynman's Path Integral

To this effect, a deeper and non-local approach is the Feynman path integral. In it one takes two spacetime points A and B, and lets a free particle follow all possible paths in between these two points as an expression of its intrinsic quantum nature. As the particle does so, it spreads its presence through the wave function as described above. Of course, not all paths are equally likely in this. Those that deviate a lot from a straight-line trajectory, which is the classically expected path of a freely moving object according to Newton, should be less probable according to everyday experience.

One can use the phase factor from above to understand this when considering the action $S_{p[t]} = \int p^2/(2m)\ dt$ along some path $p$ parameterized by three spatial coordinates $[x(t), y(t), z(t)]$, with initial and final times $t_A$ and $t_B$ playing the role of points A and B. The latter follows from the fact that the parameterization $[x(t), y(t), z(t)]$ automatically assigns spatial locations. The action is simply the time integral of the kinetic energy of a particle and is also called the energy functional. It has the same units as $\frac{h}{2\pi}$, with $p^2/(2m) = \frac{1}{2}mv^2 = \frac{1}{2}m[(dx/dt)^2 + (dy/dt)^2 + (dz/dt)^2]$, for velocity $v$. So one has a phase factor $\phi_p = e^{2\pi i S_p/h}$. The sum of all these phase factors is $\Sigma_p \phi_p$. This sum limits to an integral $\int e^{2\pi i S_{p[t]}/h}\ Dp[t]$ if one allows for small steps $D$ along all paths $p[t]$.[3]

Paths with large values of $S_{p[t]}$ lead to rapid oscillations in the phase factor $\phi_p$. Their contributions subsequently average out, thus favoring paths that have a close to stationary action, i.e., with a variation in $S_{p[t]}$ that is approximately zero. In other words, a quantum particle walks many paths simultaneously, as if it were a wave. The interference among the phase factors of these paths creates a most probable route. This elegant approach, first suggested by Richard Feynman, brings the fundamental role played by paths to the forefront using quantum superposition.

However, there is an essential problem here. The path integral approach assumes that Nature is able to provide a multitude of paths in the first place, so uninhibited by any measurement uncertainty as discussed above. This is not obvious of course. After all, paths through spacetime are not tangible like

[3]If forces act on the particle, then one can simply add an appropriate interaction potential to the free action.

matter, but nonetheless they are real entities that allow particles to travel. As such, it somehow must be possible to choose four-space paths despite Nature's fundamentally limited measurement accuracy.

Put differently, if one is to build a quantum spacetime of paths, which are themselves non-local objects, then one must be able to select and distinguish paths globally without the ability to perform precise measurements locally. How then can Nature assess any differences or similarities among paths at all if accurate measures are not available to define paths a priori? That is something to carefully explore further in the next chapter.

# Chapter 3

# Topology to Express the Quantum Uncertainty of Space and Time

From the previous chapter the central idea is that many paths together express quantum phenomena. In this, observer and observed are entangled and jointly define reality through a multitude of paths that are both distinct and equivalent. Furthermore, recall that the fabric of spacetime itself is very uncertain as well, due to fundamental measurement uncertainty. It is therefore never possible to measure, beyond any doubt, that an object takes the same path between two locations upon repeated travel, be it according to an observer like you or me or even Nature itself. Hence, an object traveling repeatedly from A to B must be able to do so along a different route each time. Analogously, two different entities traveling from A to B can always do so along paths that are distinct from each other.

In other words, the unique identity of a single path cannot exist in the presence of quantum uncertainty. This change in perspective, although subtle, is far-reaching. The fact that no one can measure two traveled routes to be exactly the same, then implies that different routes between two locations can always exist. i.e., the presence of a universal measuring uncertainty allows particle trajectories between any four-space points A and B to be intrinsically undetermined and thereby guarantees their multiplicity.

However, having distinct routes for observers and observed still implies that Nature itself should be able to tell one route from another, to establish whether they are distinct yet equivalent. As stated above, one actually has to be able to build such different routes, like roads for cars. So the paradox remains. Even though Nature most accommodate many paths because measurement uncertainty precludes the identity of a single path, any subsequent multitude of paths must be constructed without the ability to perform precise measurements.

Inevitably, the resolution is to fully embrace this relativity of identity: the only properties of four-space paths that matter to physical reality, and which provide some measure of individualism, are the ones that transcend the very quantum undeterminism of space and time. These paths, distinct yet equivalent, are explicitly constructed in the remainder of this monograph and are used to build up quantum spacetime self-consistently.

The way forward now is to use a particular branch of mathematics, called topology. One can understand its attributes straightforwardly enough. Topology derives from the Greek words "topos" for place and "logos" for study. In this field of mathematics one asserts what the intrinsic properties of spaces and objects are if one is allowed to change their shape in a *continuous* manner, as follows.

Imagine a sheet of rubber that one may deform at will without actually tearing it. This freedom of deformation allows one to give the rubber any shape. Because changes in shape proceed without damaging the rubber, they are continuous. One can see this quite easily by drawing a line on the sheet that connects two points A and B. These two points are chosen to be the start and end point of the line and are to be kept fixed while deforming the sheet. No matter how one stretches the rubber, the line does not break into individual segments. In fact, if one draws a second separate line on the sheet, also connecting the points A and B, then one can deform the rubber continuously such that the two lines fall onto each other and become one.

At this point it is good to realize that drawing a line on the sheet of rubber requires the introduction of a coordinate system to travel by. Like a sailor plotting a course and calling it a path $p[t]$, again parameterized by spatial coordinates and time $t$. There is nothing special about coordinates, they are aids that do not affect the properties of the sheet. For example, if one considers a surface in the form of a car bonnet, then drawing different coordinate lines on it indeed does not change the bonnet's underlying shape. It merely brings it out. Furthermore, one can choose any set of coordinates that one likes and interchange among them as one sees fit.

Now draw a loop on the sheet as well. By stretching the sheet and changing its shape, the loop may morph into a circle, an ellipse or remain more irregular. In any case, a loop can also be viewed as two different paths that connect the same points A and B. Suppose in this that the sheet of rubber is without holes. If so, then any path from A to B is the same as another, for fixed start and end points. Quite so, one can make all paths between some point A and some point B fall together in a continuous manner simply because the sheet can be stretched and contracted in a continuous manner. This holds independent of different coordinate parameterizations of paths $p[t]$ between points A and B.

If one cuts out a piece of rubber along the loop, one ends up with a hole in the sheet. This is not a continuous change of course and the connective nature of the sheet has fundamentally changed. It is impossible to continuously stretch or contract the rubber such that the holed sheet becomes whole again. Now draw a loop around this hole and view it again as the combination of two paths, both starting at A and finishing at B. Contrary to the sheet without a hole, it is not possible to deform these two paths, together constituting the loop, into each other. The reason is the hole that the loop is drawn around. There simply is no way to continuously stretch or contract the rubber of the sheet such that either path from A to B ends up on the other side of the hole when the points A and B are kept fixed in the process.

The important result is that the presence of the hole has changed the *connectivity* of the sheet. In a sheet without a hole any path between arbitrary points A and B can be continuously deformed into any other. In a sheet with a hole there are paths connecting points A and B in a distinctive manner: those that do and those that do not loop around the hole in the sheet. Similar and more elaborate possibilities would result if one cuts out more pieces, ending up with a sheet containing two, three or more holes. Even though all points in such sheets can be connected to each other by paths, not all paths can be deformed into each other by continuous transformations. In other words, these holes provide *obstructions* that allow one to distinguish paths under continuous deformation.

It is straightforward to apply these ideas to higher dimensions. In three dimensions one can think of a cube made from stretchable material, so that it can be shaped into a solid ball or even into Michelangelo's David. Of course, if one cuts out a cavity in such a three-dimensional space then any two paths comprising a loop can be deformed into each other. This is because the third dimension allows one to move a path around the cavity to any side. Clearly, the dimensionality matters here and it is good to consider this some more.

In three dimensions one also has the surface of the sphere, denoted by $S^2$. In fact, any closed surface, so an intrinsically two-dimensional space, requires a higher embedding dimension to exist at all. Just think of the surface of the Earth, with poles and an equator. This two-dimensional space requires an extra third dimension to be closed. In four dimensions, which is a bit more challenging but also more rewarding, one can imagine a hypercube which has a fourth edge perpendicular to the three orthogonal edges that meet at every vertex of a three-dimensional cube. In any case, two surfaces or two volumes or two four-spaces can be considered *topologically equivalent* in terms of the paths that they allow if one can be continuously deformed into the other, so path by path and for any connected points A and B.

# 1. Useful Mathematical and Physical Background

It is prudent to put the above notions on a firmer mathematical footing, as follows. A continuous transformation as above is called a *homeomorphism*. It is a continuous map between two topological spaces that is a bijection (so is one-to-one) and whose inverse is also continuous. This definition leads to the intuitive examples that a circle is homeomorphic to a square, a donut to a coffee mug and a pair of scissors to the three-dimensional rendition of the number 8. In this, a topological space can be many things. For the purpose of this monograph it is sufficient to view it as a continuum that allows one to perceive of intervals and regions, of connectivity between points A and B through paths, and of convergence onto a point (so that one may limit to a particular destination A or B).

Spacetime is thus so densely packed that every point in its fabric always has neighbors that are directly touching it. Therefore, if one takes a small piece of spacetime, then one finds infinitely many points comprising it. This way, Nature can construct paths that lie arbitrarily close together as well as routes that connect any two points in spacetime. The reader should realize that having a continuous fabric is not the same as having a smooth fabric or one without holes.

A topological space of dimension $n$ is called an $n$-dimensional manifold if each point in it has a local neighborhood that is homeomorphic to $n$-dimensional Euclidean space. Note that such a manifold need not resemble Euclidean space globally. To give a simple example, the surface of a sphere $S^2$ is a two-manifold but it is not globally a Euclidean space. The latter follows because $S^2$ is curved. Still, locally $S^2$ can be approximated by the Euclidean plane. In any case, a manifold is a fancy name for a space that is locally Euclidean.

Furthermore, two paths between the same points A and B are intrinsically different if they are homeomorphically distinct. That is, if there is no continuous deformation that morphs one path into the other. Just think back to the sheet of rubber with and without a hole. These two spaces carry different kinds of paths and cannot be transformed into each other in a continuous manner because there are simply different classes of paths due to the obstruction.

Of course, as one performs homeomorphisms, spacetime acquires a certain shape and this must have consequences for the distances between different points A and B. One might then think that everything is uncertain when topology rather than geometry is considered for spacetime. This is not so. In fact, various aspects of the relativity of space and time, as well as curvature that describes gravity, are intimately tied to both the connectivity and shape of spacetime. This is explored in the next subsections.

## 1.1. The Constancy of the Speed of Light

From now on, the discussion is always in four dimensions, unless indicated otherwise. So spacetime is a four-dimensional world with three spatial dimensions and one temporal dimension. For this purpose, one may introduce coordinates $x$, $y$, $z$ (for space) and $t$ (for time). Such a properly continuous and path connected spacetime can be endowed with a specific *metric*. This is because four-space need not be flat, but can be curved in many different ways under continuous transformations (so homeomorphisms). The metric allows one to measure distances between any two points connected by a path. In this, the distance from A to B is the same as from B to A of course. Most importantly, the metric on a four-dimensional manifold has a signature, related to the way in which the metric combines small changes in coordinates, so $dx$, $dy$, $dz$ and $dt$, into a distance measure $ds$. For Euclidean space, one has that the square of the distance increment is given by $ds^2 = c^2dt^2+dx^2+dy^2+dz^2$. This is similar to Pythagoras, but in four rather than two dimensions. The signature then comprises the signs of the metric elements, so it is $++++$ in Euclidean space because all coordinates have the same positive sign.

Furthermore, in chapter 1 it is already pointed out that the speed of light in vacuum ($c \approx 300,000$ km s$^{-1}$) is an invariant for all observers, so in any coordinate system. This means that one can always multiply with $c$ to convert a time interval $dt$ into a distance measure $cdt$, as is done in $ds^2$ above. It is important to realize that the constancy of the speed of light puts space and time coordinates on equal footing. More importantly, though, this also has far-reaching consequences for how observers perceive time, as follows.

This constancy of the speed of light in vacuum is an empirical fact, going back to the Michelson and Morley experiment of more than a century ago. In addition to this, the speed of light in vacuum is also the maximum speed at which information can and must travel in our universe. If there is some medium through which an electro-magnetic wave travels, then this propagation speed can by less than $c$. However, if some other particle carries a causal signal then it still does so at a speed that never exceeds the value of $c$ in vacuum. The existence of a constant maximum speed at which all information (so not only in the form of electro-magnetic radiation) travels has intuitive and logical basis. After all, in a causal universe one expects that information regarding an event reaches an observer with a speed that is independent of the relative motion between observer and observed. If this were not the case, then one can imagine a path between observer and observed for which the relative motion determines this causal signal speed.

That is an undesirable situation. Different coordinate systems, so defined by paths that carry observers and observed with a constant velocity difference, would then represent different physical laws. Surely, the latter constitutes a poorly defined universe. Rather, one follows Einstein's famous relativity principle: the relative motion between observers and observed should be of no consequence to natural laws, and these should hold irrespectively.

Indeed, one may perform an experiment with some apparatus (the observed) at home and then repeat the observation after putting the apparatus on a train moving at constant speed. Obviously, the path of the train, which signifies relative motion with respect to one's home, should not matter for the natural laws determining the experiment's outcome and *communicating* this at the speed $c$. Consequently, $c$ should be a true invariant for all observers and observed, along all paths and their coordinate parameterizations.

Of course, when $c$ is constant, then the spatial coordinates $x$, $y$, $z$ and the time coordinate $t$ must become mixed in such a way that it is possible for every observer to assess whether a causal signal produced by some event has traveled far enough to reach him/her. This can be facilitated *only* by a distance increment squared given by $ds^2 = -c^2dt^2 + dx^2 + dy^2 + dz^2$ with signature $-+++$. With such a measure of distance one speaks of a Minkowski spacetime. A signature $+---$ is fine as well of course, as long as space and time behave in an opposite sense. It is good to ponder this distinct yet equivalent behavior of space and time coordinates some more.

The new expression for $ds^2$ has the property that it can be less than, equal to, and larger than zero. Obviously, $ds^2 < 0$ means that a causal signal travels $cdt$ and that this is further than $(dx^2 + dy^2 + dz^2)^{1/2}$. One calls this a time-like separation of events because observer and observed can talk to each other. Conversely, $ds^2 > 0$ means that a causal signal cannot reach such a region of spacetime at the speed of light. This is a space-like separation between events. Finally, a light-like situation is provided by $ds^2 = 0$, corresponding to the maximal and optimal case where events are maximally separated but observer and observed are still in causal contact.

Upon division of $ds^2$ by $c^2$ one obtains a time squared. Hence, for small changes in all coordinates one can write for the change in the so-called proper time $\tau$ that $d\tau = (-ds^2)^{1/2}/c = (dt^2 - dx^2/c^2 - dy^2/c^2 - dz^2/c^2)^{1/2}$, which is positive and real for time-like separations. It is important to appreciate that proper time reflects the elapsed time on a clock that follows a path *connecting* two events. Upon division by $dt$ one gets a ratio between proper time and coordinate time that scales with the so-called $\gamma$ factor, $\gamma = (1 - v^2/c^2)^{-1/2}$ for the relative speed squared given by $v^2 = (dx/dt)^2 + (dy/dt)^2 + (dz/dt)^2$. Indeed, time is relative when observers move with respect to each other and $d\tau = \gamma^{-1}dt$.

The latter occurs because causal signals travel at the speed of light, independent of relative motion, through the way that Minkowski spacetime is constructed, so by $ds^2 = -c^2dt^2 + dx^2 + dy^2 + dz^2$. To bring this out consider motion along the $x$-axis only, so that a light wave obeys $x = ct$ (and $ds^2 = 0$). Two observers in relative motion use coordinates ($x$, $t$) and ($x' = x + dx$, $t' = t + dt$), where the primed coordinates indicate motion of the primed observer with respect to the unprimed observer, over a separation ($dx$, $dt$). This is just to give either a label. Then from $ds^2 = 0$ one finds $dx = cdt$ and $x' = x + dx = c(t + dt) = ct'$. Indeed, $x' = ct'$ (and $ds'^2 = 0$) holds in the primed coordinate system and $c$ is an invariant to all observers in Minkowski spacetime.

There are two more important things to learn here. First, observers can only agree on proper time $\tau$ and not on coordinate time $t$. Hence, one has a relativity of simultaneity because events that occur in unison in one coordinate system need not do so in another, given that there is no notion of absolute time. This means that two observers that move close to the speed of light with respective to each other, so with $\gamma >> 1$, see each other's clock tick much slower. Consequently, the aging of fast moving observers slows down as well. In a very real sense one can live longer by traveling close to the speed of light. Our technology does not yet allow this, though. Similarly, relative motion also leads to different observer perceptions of what the length of an object actually is, because relative duration implies relative distance.

Second, whatever positive functions one multiplies the coordinates $x$, $y$, $z$ and $t$ with, the relative signs do not change in the expression for $d\tau$. e.g., one can write $ds^2 = -c^2 d\tau^2 = -c^2 g_{00} dt^2 + g_{11} dx^2 + g_{22} dy^2 + g_{33} dz^2$ for so-called positive metric elements $g_{\mu\mu} > 0$, with $\mu = 0, 1, 2, 3$. The $g_{\mu\mu}$ functions stretch and contract the coordinate directions and correspond to $t$, $x$, $y$, $z$ respectively for $\mu = 0, 1, 2, 3$. These metric functions therefore change the distances between spacetime points, but leave the underlying Minkowski structure unaffected. Hence, the metric signature is independent of the geometry of spacetime and the motions of observers along paths. It is therefore a topological quantity. Indeed, since Einstein's special relativity derives from having a Minkowski spacetime, with causally connected observers, it too follows from topological considerations alone. Changing the shape of spacetime, as if it were rubber, does not affect this special property of our universe and is the next topic of interest on the way to quantum uncertainty.

## 1.2. Curved Spacetime

A curved manifold with a metric that locally looks like flat Minkowski spacetime is called a Lorentzian manifold or a pseudo-Riemannian manifold. These are names to rightly honor the scientists that introduced them and one should not be intimidated by them, although one may receive odd looks when mentioning such terms at a cocktail party. A pseudo-Riemannian manifold is simply a spacetime with curvature that also obeys Einstein's relativity principle. In terms of paths, such a connected manifold is a proper metric spacetime because one can define the distance between any two points by considering all possible paths between them, as follows.

One can choose the stationary action paths through curved spacetime, the so-called *geodesics*, from among the collection of all paths $\{p\}$. Recall in this the action $S_{p[t]} = \int p^2/(2m)\, dt = \int \frac{1}{2}mv^2\, dt = \int \frac{1}{2}m[(dx/dt)^2 + (dy/dt)^2 + (dz/dt)^2]\, dt$, as described above. Varying the action using standard calculus yields an acceleration $a = dv/dt$ that vanishes locally along the stationary action path. Such a geodesic is the relevant route: a freely moving observer following it detects no acceleration of particles in his/her local neigborhood along any stretch of the path.

Indeed, any pseudo-Riemannian curved manifold is locally flat with a Minkowski signature and a particle should persist in its motion if no acceleration acts on it according to Newton. Patching all these path segments together yields the geodesic between two points, or spacetime events in a causal sense. The length of this geodesic is then stationary as one introduces small variations $\delta S$ in the action along the path.

The metric coefficients introduced above can be further generalized to $g_{\mu\nu}$ for $\mu$ and $\nu$ independently equal to $0, 1, 2, 3$. So that there are 16 components in total in four dimensions. The general distance between any two points separated by intervals $dx^\mu$ is then $ds^2 = g_{\mu\nu}dx^\mu dx^\nu$. Repeated indices $\mu$ and $\nu$ are to be summed over from 0 to 3 by the so-called Einstein summation convention. Hence, $ds^2 = g_{00}dtdt + g_{01}dtdx + g_{02}dtdy + g_{03}dtdz + g_{10}dxdt + g_{11}dxdx + \ldots$. Because the distance from A to B is the same as from B to A along the same path, one has that $g_{\mu\nu} = g_{\nu\mu}$, and 10 independent components remain. Those familiar with them, may recognize that matrices form a natural representation of metrics, with coordinates in the form of four-vectors.

Also, think again of the stationary action $\delta S_{p[t]} = 0$. It is clear now that changes in it, when perceived through distance intervals $d\tau$, imply second order time derivatives. This is because geodesics derive from vanishing acceleration locally. The second order derivatives of the metric functions $g_{\mu\mu}$ therefore define the curvature, i.e., the shape, of spacetime. After all, the coordinates $x^{\mu} = (t, x, y, z)$ are arbitrary and can be varied along a local stretch of path through coordinate changes $dx'^{\mu} = (\partial x'^{\mu}/\partial x^{\nu})\, dx^{\nu}$, with the prime indicating some different coordinate system $(t', x', y'z')$ and using the Einstein summation convention. The subsequent first order variations in $g_{\mu\nu}(t, x, y, z)$ are therefore not pertinent for local motion.

Hence, the shape of spacetime must derive from the change in the change of the metric coefficients. It is not necessary for this monograph to write down the curvature explicitly in terms of $g_{\mu\nu}$ (as first derived by Riemann in four dimensions). One should simply understand that it is a geometric quantity that exists through second order derivatives of the metric, and specifies the geometry of spacetime when one performs continuous transformations of its fabric. Equivalently, those familiar with Maxwell's theory of electromagnetism may know that it also comprises second order derivatives, but then of the electromagnetic potentials. This similarity was Einstein's own guidance in the correct mathematical formulation of general relativity.

Finally, a so-called Lorentz invariant is a quantity that all observers can agree on, whatever their coordinate system or the shape of spacetime. Examples of Lorentz invariants are anything that one can one count, like the number of cows in a field or the proper time interval $d\tau$. Quantities like fundamental constants (e.g., $c$, $h$ and $G$) are also Lorentz invariants. Consequently, so are $l_P$, $t_P$ and $m_P$. Hence, the numbers describing the measurement uncertainty of quantum spacetime are themselves well defined.

All this being said, the goal of this monograph is to actually construct quantum spacetime using paths. To this effect, it is important to first be fully aware of what homeomorphically distinct paths are all about when it comes to quantum four-space phenomenology. Hence, below follow thought experiments that tie measurement uncertainty to the connectivity and shape of spacetime. i.e., the goal is to relate quantum mechanics to the mathematics of topology and geometry.

## 2. Distinct Paths Without Precise Measurements

With these mathematical excursions behind us, it is time to return to quantum uncertainty. Imagine tying a long piece of rope to starting point A (a pole of some sort) at one end, and attaching the other end to a person hiking through nature. As this person travels, he/she will leave a rope trail behind. Upon arrival at B, one may detach the rope from the person and tie it down. As one repeats this trail making between fixed points A and B, one can make different rope paths depending on the routes taken.

Now suppose one pulls hard at the ends of all the ropes at point B. One can then pull the ropes tight, but surely there will be many that get stuck around obstructions like trees, bushes, rocks, etc. In other words, without measuring how much rope was needed or where ropes were placed precisely, one even pulled at the ropes after all, it is still obvious that certain paths are different from others because they get stuck behind different obstacles.

So, provided one knows something about these obstructions, and has a rough idea about the points A and B, this deformation of the rope paths is like *distinction without precise measurements*. This is a very important result. If there are obstructions along a route, then it is possible to distinguish some of the paths simply by noting how different obstructions get in the way, without the need to know where the paths are exactly. Therefore, given that Nature cannot perform precise measurements, it has to create obstructions in order to distinguish paths...

One can make a further important step. Rather than using actual ropes and rocks, Nature can be more sophisticated and use spacetime itself to distinguish paths and build such obstructions. If Nature marks down, with its limited accuracy $l_P$, along which coordinates (locations plus times) a particle travels, then this list of numbers, the route information, replaces the rope. One can best think of marking a route on a map, like a geodesic. The same can be done for all other routes. Now recall that spacetime, which is the ultimate map to draw a route on, is like rubber. The drawn paths can then be moved around by stretching and pulling the rubber. Conversely, the obstructions are places where the rubber contains holes as above, or other higher-dimensional obstructions discussed below. Most importantly, obstructions in the fabric of four-space allow for path distinction without precise measurements, whatever spacetime geometry (so shape) one might have.

## 2.1. The Three-Torus

At this time, it is good to introduce a special space that contains obstructions and that can be used as a building block of quantum spacetime. It is the three-torus. To understand its nature, one can start one dimension lower. Think first of the two-dimensional surface of a donut. Then travel, as a very flat hiker, with ropes along the loops in the donut's surface. As one traverses around the central hole a closed path results. Similarly, one can perform a round-trip along the other loop that the donut's surface contains. Again, one runs into obstructions if the ropes are pulled tight.

One thus sees that a very good method to create immaterial obstructions in three dimensions, is to take something like a sheet, to bend it through the third dimension, and to glue the sides onto each other. For a sheet of rubber, one first gets a cylinder (gluing once) and then the surface of a donut (gluing twice), with two internal loops for ropes to get stuck around. Such loops are referred to as $S^1$ and they are basically deformable circles, like the equator of the Earth as one flexes its surface. In this, it is important to realize that the double-looped surface of a donut is a two-dimensional manifold that must be embedded in three dimensions to exist. The latter because the two gluing procedures require one to bend the sheet of rubber into the third dimension to close the loops.

To represent spacetime, four dimensions are to be considered of course. Therefore, one needs to use the fourth dimension of time to build obstacles in four-space. Imagine then taking the front and back of a solid, so three-dimensional, cube and stretching it through the fourth dimension back onto itself. Then repeat this procedure for the top and bottom as well as left and right faces. In effect, one has created three closed loops that carry any observer through the fourth dimension. This is similar to the exercise of gluing the edges of a sheet together, but in one dimension higher. It is not easy to imagine this construction and thinking of the donut's surface as an analogy is fine too. Still, one must realize that time itself is used now to create obstructions...

Such a three-way identified cube is called a three-torus $T^3 = S^1 \times S^1 \times S^1$ where the product of three $S^1$ loops reflects multiple *identifications* through the fourth dimension. This so-called topological identification of regions through a higher dimension requires an effort of abstract thinking in four-space, but the reader hopefully enjoys the time pursuing it. The three-torus is one of the two fundamental constituents of spacetime. The other is the wormhole, which is discussed further below in this monograph. It is crucial to realize that a three-dimensional object like the three-torus can only exist when the embedding dimension is four. That is, just like the surface of a sphere requires three dimensions to be embedded, so do the three loops in $T^3$ need the fourth dimension of time to be realized. Hence, the three-torus is a three-dimensional space that possesses connections running through the fourth dimension of time. This is a special object indeed. It is good then to obtain a clear image of what the three-torus provides in terms of quantum spacetime paths.

Suppose that one walks through a three-torus, say from the front to the back. As one reaches the back side, and steps through it, one reappears at the front side. As with the surface of a donut, which is also called a two-torus, one loops around through a higher dimension. One can therefore best view the three-torus as providing portals along different spatial directions but through time itself. Suppose now that an observer has a rope tied behind him/her and makes a round-trip around one of the loops. Clearly, such a traveler sees his/her rope become obstructed when he/she pulls it at both ends, even though it can be a geodesic along every local stretch of the way.

Furthermore, any path that a traveler follows actually constitutes a history. It therefore requires a sense of time and length. However, quantum uncertainty assigns uncertain moments and locations to the topological identifications that constitute the loops. Sure enough though, and even in the presence of these uncertain durations and distances in the form of local fluctuations, the traveler's round-trip is unambiguously detectable because his path is literally stuck in *multiply connected* four-space. This is the beauty of a spacetime with multiple connections carried by obstructions through the fourth dimension; it embodies quantum uncertainty through its many paths that observers and observed can walk collectively.

## 2.2. The Lattice of Three-Tori

Of course, the above only pertains to a single three-torus and one wants to build a proper quantum spacetime for an arbitrary size and age of our cosmos. To construct three-tori throughout spacetime, as a result of measurement uncertainty, one can imagine taking all three-dimensional cubes in locally Euclidean spacetime and turning them into three-tori by topological identifications. i.e., by stretching and gluing their faces onto each other through time.

Naturally, there is the question across what durations and distances these four-space connections exist. Given that multiply connected spacetime expresses quantum uncertainty, the latter must be described quantitatively by the only intrinsic measures that our universe possesses, the Planck time and Planck length introduced before. Each three-torus should then have identified edges with a length of about $l_P$ and a temporal size of about $t_P = l_P/c$, as Nature's self-imposed accuracy limits. This way, there are built-in obstructions in the form of tiny loops, which are uncertainly placed in four-space, and that ropes can get stuck around through time.

Still, doing this for all of spacetime, one ends up with many three-torus "islands" embedded in four-space. One also needs a procedure to attach all these tiny cubes-turned-into-three-tori to each other and build up a properly path connected spacetime. i.e., every Planck scale piece of four-space can be turned into a three-torus, but all such three-tori need to be connected to each other as well for paths to exist between any macroscopically separated points A and B. This sought after global gluing procedure is mathematically called a direct sum, and it works as follows.

To be able to actually travel from A to B, one must connect each tiny three-torus to its neighbors. To this effect, one can first imagine beams that run from each of the six cube faces to the opposite faces of next door three-tori, realizing that both beams and faces are respectively embedded and identified in four dimensions. Along the beams, one can go from one cube to the next, as if no $S^1$ loops exist. Conversely, as one passes through the part of a cube face that is topologically identified, one enters a temporal loop, so a portal connecting different histories (i.e., realities).

To imagine this in four dimensions is again challenging. As a three-dimensional analogy, one can glue many donuts together with dough, in a planar configuration, and then look at the resulting surface. There are now many paths that are distinct under continuous deformation of the pasted together donuts, as they loop in various ways around the many holes in the multiply connected surface embedded in three dimensions. Indeed, between any two points A and B many routes, so histories comprising a multitude of events, exist as an observer travels along the connected $S^1$ loops.

In four dimensions, this connected sum is performed by removing a solid ball, so a three-dimensional ball (usually called a three-ball), from two three-tori. The two resulting manifolds are subsequently glued together across their spherical boundaries (so $S^2$ surfaces) that are created by the removals. When one does this for all three-tori, then a *lattice* of connecting beams results that has a three-torus at each of its crossings. This lattice of three-tori embedded in four dimensions forms the backbone of quantum spacetime and is the topological expression of the quantum phenomena encountered so far in this monograph.

Again, when applying the rope trick and using continuous deformation of paths, the rope can get stuck around many three-torus loops through time. There are a huge number of these in a cubic cm of space, given that the Planck length is only $\sim 10^{-33}$ cm. Yes, that means about $10^{99}$ of them in one cm$^3$, almost a googol. Hence, there are very many topologically distinct paths through four-space. At the same time, one can travel along the beams of the lattice to navigate around the loops, somewhat oblivious to the hidden secrets of the lattice of three-tori.

As a bonus, the $S^1$ loop can also be viewed as a circle in the complex plane. The mathematical action of conjugation, introduced before, then yields the squared modulus $\psi\psi^*$ through flipping the sign of $i$. Conversely, the physical notion of a closed path forms the origin of the square as well, because it allows observers and observed to exchange their roles along four-space paths. The omnipresence of $S^1$ loops thus allows every such observer-observed pair to mirror each other's phase factor.

In all, one has constructed many different paths, between any two spacetime points A and B, that appear invisible to the macroscopic world and that are identifiable as distinct routes by the way they get stuck around tiny Planck scale loops as spacetime is stretched and pulled like rubber. Therefore, one has succeeded in building many topologically distinct routes without precise measurements. The resulting lattice of three-tori is a multiply connected spacetime in a strict mathematical sense because homeomorphisms (continuous transformations) cannot make the embedded loops go away.

Feel free to ponder this for a while. Realize also that these different quantum routes, despite the small values of $l_P$ and $t_P$, exist on all scales. After all, two macroscopic paths, even traversing light years, can already be made distinct if one path picks up just a single Planck scale loop and the other route does not. As such, tiny topological differences can have profound everyday consequences. This is precisely what one would expect in a universe where homeomorphisms allow for the quantum equivalence of all shapes and sizes. These many paths are traveled, and the multiple histories enjoyed, by the entangled observers and observed that occupy the universe. Hence, using the lattice of three-tori, quantum spacetime can be perceived as a collection of histories that are fully connected through all of time.

Furthermore, the lattice of three-tori embedded in four dimensions allows one to view Feynman's path integral formulation of quantum mechanics from a different perspective, as follows. It is the path aspect that is crucial in Feynman's approach to quantum phenomena, allowing phase factors to be formulated in terms of actions along given routes. Because there are many topological distinct paths between any points A and B, matter travels along all of them. It is therefore only the cumulative behavior of interference among a multitude of routes, now deeply routed in spacetime itself, that embodies quantum behavior. Consequently, it is the lattice of three-tori that is from hereon on taken as the quantum spacetime expression of Feynman's path integral. In other words, quantum mechanics can be formulated through actions along paths simply because the spacetime fabric comprises them by its multiply connected (so temporally obstructed) nature.

In this, one may worry that using the fourth dimension of time to introduce quantum obstructions into the spacetime continuum leads to problems with cause and effect. Even though causality can be affected, such is the price that one has to pay upfront. While Einstein's deep ideas about relativity imply different rates at which clocks run, as discussed above for motion and also in the next chapter for gravity, the lattice of three-tori implies uncertainties in the temporal ordering of events... There is no way around this.

Indeed, because the speed of light $c$ is constant, according to special relativity, any measuring uncertainty $\delta l$ in cm corresponds to one in time of $\delta l/c$ s. Given that a path through spacetime is just a string of events, so a history, the placement of events in time cannot be exact either. Fortunately, the same $S^1$ loops that connect all entities through time by means of topological identifications, and thereby entangle observers and observed, also constitute the temporal obstructions that allow Nature to distinguish paths.

In fact, because these topological identifications of the lattice of three-tori are closed loops, which can adapt to any spacetime shape under the global actions of homeomorphisms (so quantum measurement uncertainty), they in effect *induce* the very reality that observers and observed experience individually. i.e., $S^1$ loops, so portals through time, constitute pairs of paths of observers and observed that may always *exchange* their roles. Consequently, without there being a chicken and egg problem, these distinct yet equivalent observers and observed follow from the $S^1$ loops and the latter can therefore be seen as the inductors of intrinsically discrete quantum realities.

Indeed, because entire histories can be distinguished topologically by $S^1$ loop obstructions, so can most of the events experienced by observers and observed that occur along them. Hence, despite the abstract impression that $S^1$ loops in the lattice of three-tori make, their fundamental (and everyday) property is to simply allow the exchange of observers and observed. This in turn allows for both the persistence and alteration of such a jointly carried quantum world. As such, these inductors of reality truly are the backbone of tangible existence.

Nevertheless, unlike a spacetime without a lattice of three-tori, causality is no longer strictly guaranteed globally since the order of events can be shifted as Nature continuously deforms paths in a somewhat undetermined manner. That is, as one pulls the thought experiment ropes and stretches the spacetime rubber, one may shift the history of one observer or observed with respect to another. The keyword here that saves one from chaos is globally. As discussed before, the quantum interaction between observer and observed picks out an entire four-space history that is locally causal, yet comprises globally entangled events.

Consequently, without gravitational wormholes known in popular literature and discussed in chapter 5, and without Einstein's theories of relativity being wrong in any way, one may conceive of time travel as being allowed when there is a fundamental and universal measurement uncertainty rooted in spacetime itself. Of course, these global temporal connections between different events, through the lattice of three-tori, pertain strictly in a quantum sense and are discussed further once quantum spacetime is thoroughly constructed (see also chapter 10 on the multiverse).

Also, as Nature builds the lattice of three-tori, the placement of Planck size three-tori is itself uncertain by about $\delta l = l_P$ in space. That is, when a particle starts inside an arbitrary three-torus and travels along the beams of the lattice (top-bottom, front-back, and left-right) by about a Planck length, then it moves into one of the six neighboring three-tori. However, within the measurement uncertainty $l_P$, these neighbors can be mistaken by the particle for the central three-torus. So, from the perspective of the particle, or any observer and observed, the three-tori form groups of seven. These septuplets of three-tori are thus like identical siblings that all live in about the same place and time, but not exactly so. i.e., one septuplet member is just as good as any other because they are equivalent. Hence, if one mixes them up, together with the paths that they distinguish, then one has a distinct yet equivalent history because no one can know, not even Nature itself. This implies that there is quite some redundancy in the way that Nature assesses the obstructions provided to traveling particles by the lattice of three-tori and therefore in how it keeps track of complete four-space histories of observers and observed. It is important to keep this in the back of one's mind, particularly given the very large number of three-tori that are present in any given $\text{cm}^{-3}$, and this for any time.

In fact, at this point a somewhat philosophical comment is in order: whatever one can think of, can exist. That is, even though mathematics is a formal language, derived from definitions and logic, and physics is a mostly intuitive endeavor, they are both expressions of human thought. This while those of us pursuing ideas about the universe are all part of it. Hence, it would seem quite implausible that anyone can think of something beyond the boundaries and proper fundamentals of one's own existence. Consequently, the multiply connected lattice of three-tori conceived of above is worth pursuing for all its merits, whatever strange perspectives it may yield, provided one's reasonings are *self-consistent*. To have the proper luggage for such a journey, and to build upon the lattice of three-tori, the focus in the next chapter is on gravity.

# Chapter 4

# Aspects of General Relativity

Gravity is one of the four fundamental forces in Nature, together with the electromagnetic, the weak and the strong force, and it is the most important force for the formation of stars, black holes and galaxies in our universe. One of the beautiful aspects of the gravitational force lies in its connection to geometry, so the shape of spacetime. To appreciate this, it is prudent to introduce Einstein's original thought experiment, and to study it from a new (quantum path) perspective.

## 1. The Equivalence Principle

Imagine an observer traveling through a spacetime that undergoes homeomorphisms, so continuous deformations. In this ever-changing landscape, the observer experiences curvature depending on the particular shape that his world takes. Just think back of the metric above, with its second order changes setting the ambient geometry. As the observer follows a path from A to B, without performing any action or being subject to any action, he/she notices that he/she does not deviate from a geodesic. This is what one expects from the Feynman path integral formulation in terms of a stationary action, i.e., a freely moving observer who detects no acceleration of particles in his/her local neigborhood.

Conversely, the fact that the observer's world is curved, so non-Euclidean in a global sense, suggests a physical significance besides quantum measurement uncertainty, as follows. A free-falling object that simply traces the natural shape of spacetime, so along a geodesic, does not feel any acceleration if one ignores air friction. Furthermore, an observer keeping himself/herself fixed, like on the surface of the Earth, does feel a constant pull in response to his/her effort. After all, action in a mechanical sense is required to maintain oneself upright. Such a stationary observer also asserts that freely moving objects do accelerate with respect to him/her. That is, depending on one's frame of reference there is a particular experience in terms of acceleration. Note how this differs from special relativity, which deals with constant relative motion.

All this implies that the shape of the observer's spacetime is equivalent to a force that behaves as a local acceleration field.[1] The inevitable conclusion is then that spacetime curvature expresses gravity *if* the inertial and gravitational mass of any particle are the same. That is, if there is one kind of mass that pertains to a particle's change in momentum under both inertial and gravitational forces. This is what Einstein concluded when considering his famous elevator thought experiment. Hence, this is a good time to present it.

Imagine an observer standing in an elevator suspended above the ground. He/she feels the pull of gravity as a local acceleration. Suppose now that there are no windows in the elevator to verify one's whereabouts. Under those circumstances, the observer would enjoy the same experience if the box of the elevator is somewhere in free space and there is a rocket that accelerates it at the same rate as the Earth's gravitational field does. Of course, in the Earth's case the observer feels a force equal to the acceleration times the gravitational mass. In the free space case the force is proportional to the inertial mass. Given that these two masses are empirically found to be equal, one can always express the effect of gravity as motion relative to an accelerated frame of reference.

Although (deceivingly) simple, this conclusion has far-reaching consequences because the relativity of reference frames is what spacetime is all about. Indeed, observers that enjoy constant relative motion as in special relativity are described globally by Minkowski spacetime with its $-+++$ signature, a topological invariant. This while observers that experience gravitational accelerations occupy a curved spacetime, carried by connectivity preserving homeomorphisms and with Minkowski spacetime pertaining only to local regions.

The notion that there is a complete physical equivalence between the local strength of gravity and the local acceleration of a reference frame is called the

[1] A field is a function that has a value in every spacetime point.

equivalence principle. Movement along a geodesic in a curved spacetime then equates to the force of gravity. Indeed, everyday experience tells us that free-fall, like a jump from a building onto a cushion, feels like one is weightless. Hence, a free-falling observer does not experience any gravitational effects, unless the gravitational field is non-uniform due to tidal forces (this formulation is called the weak equivalence principle). Of course, when one considers a small enough spacetime region, tidal forces can be made negligibly small.

Furthermore, two such skydivers close together enjoy the same experience. So every free-falling observer is oblivious to a sense of acceleration and perceives local spacetime as flat and Minkowski. But if every free-falling observer perceives spacetime as locally flat, then spacetime cannot be flat globally if there is to be gravity. This is where curvature enters as the expression of the gravitational force. It is therefore only the *change* in acceleration, from one locally free-falling system to the other, that expresses itself in terms of spacetime curvature through the distribution of mass. Such a free-falling state is called a local inertial frame. i.e., Newton's first law of inertia applies in it and an accelerometer measures zero acceleration. In this sense, gravity is not a real force. It is more an expression of how local inertial frames, all without a real notion of gravity, are connected to each other globally.

Above it was concluded that the speed of light and the very laws of nature are independent of the relative motion of one reference system with respect to the other. For constant speeds this is usually referred to as special relativity. This coordinate system independence is now seen to also apply to uniformly accelerated frames and is called general relativity. Indeed, the laws of Nature must be independent of any choice of coordinate system, and the laws of Nature take the same form in a free-falling reference frame as they would in the absence of gravity (this formulation is called the strong equivalence principle).

Recall now that homeomorphisms express the notion that measurement uncertainty introduces the freedom of continuous deformations into multiply connected quantum spacetime. This to allow for distinction without precise measurements. With the equality of gravitational and inertial mass, any particular geometry that spacetime can enjoy, is now tied to the effect of gravity. To quantify these shapes, as a physical expression of gravity, the observer notes that the local curvature that he/she experiences is just the change from one inertial frame to the next.

To tie this to an actual acceleration $g$ in cm s$^{-2}$, just realize that this difference in local inertial frames is the distinction between two different local accelerations, so $dg/ds$ for some coordinate change $ds$ along a path.[2] Also, $dg/ds \propto G\rho$ for mass density $\rho$ according to the Gauss law of Newtonian gravity, with $G$ the gravitational constant. Because changes in acceleration express the shape of spacetime, one has that $\mathbf{R} \propto G\rho$ for the local curvature $\mathbf{R}$. Clearly, spacetime becomes more strongly curved when one compresses a larger mass into a smaller volume. As one would expect intuitively. Curvature has physical units of length$^{-2}$ or equivalently mass$^2$. This is no surprise for a force that derives from the interaction between pairs of masses. i.e., for a force whose magnitude scales as the product of two masses.

All the above again indicates that gravity is not a real force. Rather, it is a consequence of a universe that undergoes homeomorphisms, as an expression of quantum uncertainty, with a specific shape tied to a global distribution of mass density through the equivalence principle. Furthermore, because spacetime is locally flat, it is only the difference between locally flat regions that has any meaning for gravitational dynamics globally. In this, it is thus fascinating that the geometric expression of gravity, through shape, is not a priori at odds with quantum mechanics since the latter is the expression of spacetime's topology, through path connectivity.

## 2. Time and Gravity

Of course, as one curves spacetime the durations that observers measure in different reference frames can differ. This is because all four dimensions, so also time, are deformed by homeomorphisms. To make this more precise, recall the metric from above, in particular the coefficient $g_{00}$ that stretches or contracts the time coordinate. It is easy to see that $g_{00} = 1 - 2GM/(rc^2)$ for an observer at a radial distance $r$ from a point mass $M$, as follows.

Consider a light source at the top of the rocket that is accelerating at a constant value of $g$ cm s$^{-2}$ in the elevator thought experiment. The source emits photons that experience a Doppler shift as they reach the bottom of the rocket after a coordinate time $dt$. This Doppler shift is given by $\nu^{|} = \nu(1 + v/c)$ for

[2]The acceleration is actually a vector with directional components $x$, $y$ and $z$. In Cartesian coordinates one should write $dg_x/dx + dg_y/dy + dg_z/dz$.

the received frequency $\nu^{|}$, the emitted frequency $\nu$, and $v = gdt$. Furthermore, if the height of the rocket is $H$, then $dt = H/c$ and $v = gH/c$. This yields $\nu^{|} = \nu(1 + gH/c^2)$, so that $\Delta\nu/\nu = gH/c^2$.

Because of the equivalence principle the *same* effect must pertain to a gravitational field. The latter's acceleration obeys $g = -d\Phi/dr$ for the gravitational potential $\Phi$ as a function of height $r$. One has the obvious solution $\Delta\Phi = -gH$ for the difference in gravitational potential between the top and the bottom of the rocket. Therefore, photons experience a gravitational blueshift $\Delta\nu/\nu = -\Delta\Phi/c^2 > 0$.

In addition to this, the frequency $\nu$ of any electromagnetic signal is the same as the reciprocal of a time interval $T$. That is, any clock ticks at some pace and its beat frequency is what sets the passing of time. In other words, the light source in the rocket is actually a clock that is as real as one's own heart beat, with $T = 1/\nu$. Consequently, $\Delta\nu = \Delta(1/T) = -\Delta T/T^2$ and $\Delta\nu/\nu = -\Delta T/T$. It follows that clocks at the top and bottom of the rocket, or equivalently clocks separated by a radial distance $H$ in a gravitational field, obey $T^{|} = T(1 + \Delta\Phi/c^2) = (1 - gH/c^2)$ since $\Delta t/t = \Delta\Phi/c^2$. In this, $T$ is proportional to the time passed at the top of the rocket, so higher in the gravitational potential well. In general, $\Delta\Phi/c^2 << 1$ holds for a weak gravitational field like the Earth's.

Therefore, clocks tick slower when one sits deeper inside a gravitational potential, like closer to the surface of the Earth, because $T^{|} < T$. This is called time dilation and has been detected unambiguously through measurements using atomic clocks. Note that the relativity of time implies that only the ratio $T^{|}/T$ is pertinent. That is, one may freely choose a unit (like seconds, minutes or hours) to measure time in. Once such a choice is made, the ratio of measured times is subject to the effect of geometric gravity through the equivalence principle. It also follows from this thought experiment that light is deflected in a gravitational field. Imagine a light ray that travels horizontally across the rocket cabin. This path appears to curve when the rocket accelerates toward the path of the beam, so as photons move from left to right (or right to left) and perpendicular to the direction of motion. Indeed, $\mathbf{R} \neq 0$ must hold then as well.

Furthermore, for a point mass $M$ and a radial distance $r$ from it, one has that $\Delta\Phi = -GM/r$. So one finds that $d\tau^2 = (1 - GM/(rc^2))^2 dt^2 \approx (1 - 2GM/(rc^2))dt^2$, for the proper time $\tau$ and ignoring spatial effects (i.e., $dx = dy = dz = 0$). One can now read off the purely temporal metric element as

$g_{00} = 1 - 2GM/(rc^2)$. Indeed, the relativity of time can be expressed through spacetime curvature.[3] One further sees that $d\tau$ goes to zero for a radius equal to $2GM/c^2$. As the reader undoubtedly realizes, such a radius is special because proper time appears to slow to a complete halt. Conversely, if proper time is to be non-zero for $g_{00} = 0$ then coordinate time must limit to infinity. This highly unusual (but tantalizing) situation is indicative of the existence of black holes and wormholes, and is explored further in the next chapter.

In practice, though, time dilation is a modest effect because $\Delta\Phi/c^2$ is typically quite small, so much less than unity. Consequently, the longevity that one may gain by spending more time on the ground than on top of a mountain is almost negligible for planet Earth. On average, it is not more than a few milliseconds over a human lifetime of 80 years, a height $H \approx 9$ km (like Mount Everest) above ground, and for the Earth's gravitational acceleration of about 10 m s$^{-2}$. With the basic principles of general relativity in place, it is time to dig a bit deeper and to consider mass itself.

## 3. Mach's Principle

The attentive reader may have noticed that individual particles are always assumed to have a well defined mass to yield the density $\rho$ or cumulative mass $M$. Even though inertial and gravitational mass are measured to be equal, so it is an empirical fact, one may wonder how inertia can be measured at all if there is no absolute standard of rest for motion in a dynamical spacetime. Of course, such assessments of inertia are performed with respect to some fiducial measure only, say a standard kilogram. Still, it is undeniable that a notion of mass pertains to any observer in a local frame of reference. This while the collective of all observers and observed, so the entire universe, must somehow be responsible for every physical phenomenon, including the inertia of particles.

This led Mach to a statement of the effect that matter at distant locations influences inertia nearby. That is, the inertia of local matter is the result of the global interaction between bodies throughout the universe. The latter thus define, as a collective, a property that local observers perceive as inertia. The key concept here is global, i.e., the sum of all things determines its individual

[3]Note that Einstein's classic and popular book "Relativity" is still, even though a hundred years old, a pleasure to read for those interested in the theory of relativity.

elements. So this principle of Mach, besides inertia, assigns special meaning to the whole of all things determining the properties of its constituents.

Realize then that general relativity, through the equivalence principle, is a local theory and all particles in it are endowed with an intrinsic mass a priori. As such, Einstein gravity does not fully live up to Mach's principle, notwithstanding general relativity's overwhelming experimental success. Interestingly, the global distribution of matter that Mach and Einstein thought of acquires a new status in the context of paths through quantum spacetime. After all, the lattice of three-tori embodies the fact that particles can enjoy interference effects through a multitude of four-space histories, so can act as a collective in a truly global sense.

With this distinction between local and global, there is more to learn from Einstein's beautiful elevator thought experiment. To appreciate this one should realize that the above explorations show that general relativity is a local geometric theory, dealing with the shape of spacetime and leaving room for any global topological information on the connectivity of the four-space fabric. This while quantum phenomena derive from a spacetime built up by a multitude of paths. A more explicit link between these two perspectives is what one seeks.

## 4. The Role of Information in Gravity

To this effect, there is a fundamental quantum connection that one can establish between observer and observed, and one that is already present within general relativity in an implicit manner, through *information*. When brought out, this connection constitutes an important step toward a marriage between quantum mechanics and general relativity because it shows the relevance of entanglement in gravity. To achieve the above, it is good to start on the quantum side of things.

The mystery of quantum mechanics has been explored by interference experiments between beams of light and electrons, but it can be taken a step further. If one performs the double slit experiment from above with atoms instead of electrons, then an interference pattern again appears. The advantage of atoms is that they have an internal structure since there are electrons orbiting a nucleus of protons and neutrons. With a laser tuned to a specific frequency it is possible to excite one of the many quantum states in the atoms taking part in the experiment. In fact, one can choose a quantum state of very modest energy. This to ensure that the photons comprising the laser do not change the momentum of the atoms. In this situation, the momentum of all partaking atoms is unaffected. Also, one can decide to aim the laser at only one of the two slits. The experiment then allows for the tagging of atoms going through the laser exposed slit, because they are brought into an excited state while the other atoms are not.

Of course, interference is the result of quantum particles traveling through both slits. So, with paths the root cause of the quantum mystery, labeling atoms along paths going through one of the two slits should destroy superposition. This is because the atoms can be told apart through their laser induced excitation. In other words, one performs an observation with the laser beam that distinguishes the paths going through one slit from those that go through the other. In this, one merely uses the information necessary to describe an atom's quantum state. For a simple two-level system, with a ground state and excited state, one only requires a single bit of information per atom to distinguish one slit from the other.

Indeed, it is experimentally found that such a single bit of information, attached as an identifier to one of the two slits through the excitation of atoms by a laser, is sufficient to destroy the interference pattern. It is important to realize that one really deals with just a single bit of information and not with any photon momentum in the laser beam that affects atoms mechanically through Newton's principle of action and opposite reaction.

The above empirical result testifies to the notion that any entanglement of observer and observed, as in Schrödinger's cat example discussed earlier, is the result of the global paths that exist in quantum spacetime. These distinct four-space histories are carried by the lattice of three-tori. Any a priori distinction between observer and observed is therefore pathological. Both observer and observed occupy these global paths, in a distinct yet equivalent manner and characterized by mere information. This is again a fundamental perspective to embrace.

Now consider again Einstein's elevator/rocket thought experiment, where one compares an observer in a suspended elevator on Earth to an observer that is accelerated by a rocket through empty space. The operator of the rocket that is responsible for the equivalent acceleration of the moving observer needs to be informed about the strength of the gravitational field that pertains to the suspended observer, if the operator is to **mimic** the proper experience. Without this information he/she cannot realize the thought experiment as an actual and proper physical event.

There is great significance to this assertion. It indicates that the rocket operator, who generates the gravitational field that the observer perceives, is also an observed in a quantum sense. After all, the observer in the elevator box detects the actions that the operator performs. Observer and observed are thus globally connected to each other, tangibly through a rope between rocket and elevator box, but more importantly in a topological manner through a collection of distinct yet equivalent four-space paths that jointly carry the information on (gravitational) acceleration.

Hence, global quantum information on a gravitational system, in some form, is necessary to define a physical situation that mimics gravity as a local acceleration field through observer-observed entanglement. This argument enlarges the equivalence principle with a sense of global constructability. It does not diminish the benefits of the equivalence principle in any way, since the latter derives from the local indistinguishability of physical experiences.

Put differently, one is lead to the position that the law of gravity is not imposed on individual observers and observed as an exterior rule. Rather it is a consequence of the global quantum information that is necessary to mimic local gravitational dynamics. That is, this necessary information just as much implements gravity as it induces gravity. So what the mimicking operator, the observed, needs to know (acceleration due to mass density) determines what the observer experiences (curvature), *and* vice versa. This is a true expression of observer-observed entanglement.

The required information that is pertinent here, is the one on particles since they gravitate through mass density. This is a good time then to be more precise about information. Note first that an isolated observer, so a purely non-quantum entity (defined merely as a limiting case), requires only local information to be described himself/herself. Global quantum information is needed for the matter-induced curvature of, so the operator-induced acceleration experienced in, the spacetime that all observers and observed jointly occupy. Of course, global quantum information is what distinct yet equivalent four-space paths connecting observers and observed/operators are all about. Paths carry quantum measurement uncertainty through the continuous deformation of spacetime's shape, thus providing for curvature through geodesics.

To formalize and quantify this, one can introduce local information on particles in the form of $N$ bits and global information in the form of $N'$ bits. In this, the magnitude of the acceleration that the operator needs to mimic is incidental and it only matters for him/her to know which particle properties play a role in the physical system that requires mimicking.[4] Hence, one merely needs *identifiers* that allow one to tally and distinguish local classical from global quantum processes.

The local information can then be thought of as Boolean variables that indicate with 0 and 1 whether intrinsic properties (i.e., mass, charge, spin) and four-space coordinates of particles ($x$, $y$, $z$, $t$) are required individually or as an ensemble for physical interactions (collisions) between particles. This thus requires maximally seven bits per particle. One should realize that there are likely more than $N_p \sim 10^{80}$ particles in the observable universe and that $N \leq 7N_p$ is a very large number indeed.

The global information in turn pertains to any additional correlations that exist between these particles, i.e., quantum entanglement. There should be a simple yes or no answer stored for every pair of particle properties as to whether they are correlated in some way. This leads to $N' \leq (7N_p)^2/2$ if every particle is fully correlated with every other. In this, care should be taken to assess whether an observer is part of a physical system under study or not (see immediately below). In any case, $N' >> 7N_p$ is an even larger number of bits. Nevertheless, it is important to realize that these estimates for $N$ and $N'$ are strict upper limits and that in practice much less information may be required.

Of course, Einstein's geometric gravity is a local and strictly non-quantum theory that derives from a matter field rather than individual particles. Hence, $N' = N$ for general relativity. This while the global quantum information $N'$ is precisely the observer-observed entanglement that is stored in the lattice of three-tori, so in the topologically distinct four-space paths that all quantum particles travel along. Consequently, $N' \neq N$ holds when one considers all observers and all mimicking rocket operators collectively, and thus quantum deviations from Einstein gravity can occur.

In fact, a very basic expression of global and local information is a collective of $N_s$ gravitating particles that are not strongly correlated mutually but are globally entangled with an observer $O$ (like you or me). So $N' \approx (6N_s + 1)N'_O$

[4]Indeed, on the quantum level it only matters how something happens and not what happens.

for neutral particles that require mass, spin plus four coordinates, and just a single common number for their zero charge. This while the mimicry identifier $N'_O$ expresses the global character of the quantum interaction. Clearly, one has that $N'_O = 2$ bits when observer and observed both need to be identified to form a joint physical entity. Just think of an observation where one is detecting light from a distant source.

Furthermore, if these neutral particles are *collisionless*, i.e., do not undergo local interactions through close proximity, then $N \approx 2N_s + 5$. This follows because such neutral particles require a mass and spin for their individual gravitational properties, but are otherwise defined as an ensemble by four average spacetime locations (so in a statistical sense). Put differently, if objects do not meet locally, then due to this joint isolation one does not need to know their precise whereabouts anyway. This is certainly the simplest manifestation of global information versus local information. Still, there is a significant difference between the two amounts of information.

Indeed, the ratio $N'/N$ is about six for this collisionless system, e.g., a galaxy comprised of $N_s >> 1$ stars that are neutral point particles when viewed from afar. This shows how even modest non-local information causes the strength of gravity to deviate from purely local general relativity. i.e., whatever gravitational system one considers, it is the gravitational constant $G$ that converts a measure of mass density into an amount of spacetime curvature (see above). The value of $G$ is determined through local laboratory measurements on Earth, while gravity derives from global observer-observed entanglement. Consequently, the strength of gravity must depend just as much on $G$ as on the *relative* amounts of information that are needed by the mimicking operators discussed above. After all, the ratio $N'/N$ is the only quantity that has distinctive meaning among observers and observed/operators that can always adopt each other's role. Hence, to apply $G$ correctly in a global quantum spacetime setting, one should actually use $G' = G[N'/N]$. The latter quantity need not be equal to $G$.

One should therefore always pay close attention to how local/non-quantum (collisional) and global/quantum (entangled) a system really is. Such notions depend on spatial scale of course. Just think of a gas that is being compressed from a very diffuse to a very dense state, with an observer that carries global quantum correlations and all. Similarly, a distribution of stars comprising a galaxy is an almost collisonless ensemble in practice, even though individual stars are not. Hence, as a galaxy becomes more compact, collisions/interactions occur more frequently and more local information is required. This while entanglement in gravitational systems is quite relevant for astronomically large scales, so with spatially separated observers and observed, and $N'_O = 2$. Conversely, for the Earth orbiting around the Sun, the latter is clearly an extended ball of collisional gas, with a mimicry identifier of $N'_O = 1$. Indeed, observers on Earth are an integral part of the joint gravitational Earth-Sun system, by construction. Consequently, and unlike the $N'_O = 2$ galaxy example above, $N' = N = 7N_p$ for the $N_p$ individual charged particles comprising the Earth-Sun system.[5]

The above effect of gravity driven by information amounts to the well known notion of dark matter. This is an electromagnetically invisible putative entity that expresses itself solely through gravity. Empirical evidence for the latter derives from the rotation curves of galaxies as measured in the atomic hydrogen 21 cm line (a hyperfine transition first predicted by the astronomer van de Hulst). The total required mass within the orbits of this atomic hydrogen gas, to explain its dynamics, is equivalent to about six times the luminous contribution of stars. One may perceive this implied larger mass as an invisible form of matter, hence the name dark. However, such an effect can also proceed through an effective gravitational constant $G' = G[N'/N] \approx 6G$ that implements local versus global information.

This $G'$ incorporates, even for Newtonian (non-relativistic) dynamics, what input the rocket operator needs to know in order to describe a specific tangible system and to realize an associated gravitational effect. It is prudent to explore this connection between local and global information further. After all, how can the lattice of three-tori store global quantum information and define the very reality, with gravity and all, that one perceives?

[5]Of course, viewing the Earth-Sun system from afar does yield an information correction to gravity, given by twice the ratio of the number of particles in the Earth to that in the Sun.

# 5. The Correspondence Between Geometry and Topology

A first step is to relate the lattice of three-tori to Einstein's theories of relativity in a geometric (so shape) sense. In quantum spacetime, special and general relativity remain valid because Einstein's ideas derive from the local properties of spacetime, i.e., for short distances between points A and B. Indeed, a small part of a three-torus looks exactly the same as a similarly small part of a solid cube in terms of geometry. It is therefore not possible to tell the difference unless one travels the entire cube or three-torus and distinguishes these paths globally through topological obstructions in the form of $S^1$ loops. Just think back to the rope thought experiment. Also, whatever route a particle (e.g., an electron, proton, photon or neutrino) takes, it always does so with a local speed that does not exceed that of light in vacuum, in accordance with special relativity.

Also, gravity is described in general relativity by the curvature of spacetime. Just as for the lattice of three-tori, continuity of spacetime is also intrinsic to general relativity, because the latter works with smooth geometries (so smooth shapes) that are a subclass of all continuous spacetimes. It is also important to realize again that Einstein's beautiful theory of general relativity does not use the properties of spacetime topology. This is a blessing in disguise because it allows one to extend Einstein gravity with spacetime topology without diminishing its merits.

In fact, all the beams that are used to glue the three-tori together to a four-dimensional lattice form themselves a very fine-grained mesh. These connecting beams can subsequently curve and bend according to the matter density that occupies them, and thus represent gravity as a local geometric phenomenon. As if they constitute an otherwise feature-less spacetime continuum. This even though the lattice of three-tori is globally multiply connected through the many $S^1$ loops. Again, one can have his/her cake and eat it too.

Furthermore, precisely because there are three-tori present on the Planck scale, quantum spacetime comes with additional degrees of freedom to serve $N'$. That is, global quantum entanglement information resides in the topological identifications that define each individual three-torus, the three $S^1$ loops. These loops are special, as already mentioned above, in that they are portals through the fourth dimension of time, so connect different realities. Indeed, these portals constitute temporal obstructions that induce a quantum reality of distinct yet equivalent observers and observed. The latter are entangled because their respective roles are fully exchangeable, an action that is physically facilitated by the $S^1$ loops. For each three-torus one has three such portals. So on the lattice, the number of possibilities are almost limitless as objects follow paths along curved beams (geometry) and discrete $S^1$ portals (topology) that act as reality inductors for pairs of exchangeable observers and observed.

In fact, even though the three-tori are individually tiny, as a collective they allow for loops of any size. After all, as one passes into the fourth dimension through any $S^1$, one can continue as far as one wants because all loops are connected to each other. Hence, this facilitates the distribution of global quantum information, while preserving the local causal structure of relativity theory. Indeed, even when general relativity has singled out a particular spacetime geometry, the fundamental measurement uncertainty $l_P$ maintains a multiply connected quantum world on all scales through a multitude of histories. It is only the connection of local events that causality pertains to.

Therefore, in a very real sense, time itself is multiply connected. This is the true nature of the global information $N'$. To make this more pertinent, imagine two particles each enjoying a four-space history. When these two particle paths share a single $S^1$ loop, this portal acts as a reality inductor that yields two distinct yet equivalent histories for an observer-observed pair. This loop then literally switches like a bit between these histories, and the observer and observed on either path are entangled. Just recall the two electrons in superposition from above. The mimicry identifier $N'_O$ is then the macroscopic manifestation of entanglement in terms of information. One has $N'_O = 2$, providing two distinct yet equivalent histories to define either identical particle individually, and to allow for their exchange. This quantum information is spread out globally by the very homeomorphisms, so continuous spacetime deformations, that preserve observer-observed connectivity.

## 5.1. A Few Words on Particles

Through every single three-torus, $S^1$ loops embody information on local symmetries as well. For example, imagine going around a loop time and time again as one travels from one three-torus to the next. Each such an internal loop along any spacetime path expresses the particle property of having a *single* revolution identifier. Similarly, one can do this with two loops, as follows. Suppose that one a) passes through one loop just short of closure, b) traverses the other just short of closure, and c) closes at the starting point of a) with the loose end of b). This amounts to a *double* revolution identifier, without a single revolution one present. Having both such identifiers at the same time exhausts all possibilities for a single three-torus.

Of course, a reader with some particle physics background recognizes these particle identifiers as spin 1 (bosonic) and spin $\frac{1}{2}$ (fermionic) properties, respectively. Hence, quantum uncertainty expressed through the topology of spacetime naturally yields the occurrence of bosonic and fermionic particles. In any case, these local identifiers of particle properties through the global sharing of $S^1$ loops and the bending of the beams in the lattice of three-tori nicely relate the latter's path topology to the matter induced geometry of Einstein gravity.

## 5.2. A Few More Words on Mass

One can now further appreciate the importance of a global notion of inertia. Einstein beautifully connects the inertial movements of free-falling observers to the local geometry of spacetime. This while the lattice of three-tori allows for a global multitude of quantum paths. In this, matter comprises particles with an intrinsic mass. The point of Mach's principle is to define this mass, and the dynamics of inertia in general, in a self-contained global manner.

In a quantum sense this entails that the dynamics of quantum spacetime, where paths rule particles, must somehow introduce inertia. The latter process also needs to proceed in such a manner that it can be captured by information $N$ (local particles and density) and $N'$ (their global entanglement and curvature). This inevitably calls for a collection of objects that behave as massive particles, constitute geometric objects in Einstein's sense, and act as topologically distinct yet equivalent paths which express a global notion of inertia. What can these entities be?

It is indeed possible to conceive of such objects. Solutions of general relativity, for very strong gravitational fields and thus extreme spacetime curvature, are the black hole and the wormhole, described in the next chapter. In topology one speaks of a wormhole as a handle, like the one on a coffee mug but in one dimension higher. For now one only needs to realize that black holes and wormholes act as paths between spacetime points.

In addition to this, black holes and wormholes also carry mass intrinsically. Indeed, even without any precise knowledge of what they are, one merely needs to think back to the temporal metric element $g_{00} = 1 - 2GM/(rc^2)$. The fact that $g_{00}$ can be zero introduces a characteristic radius of $2GM/c^2$ for a sphere $S^2$ with enclosed mass $M$. Conversely, this indicates that black holes and wormholes require space, time and mass for their very existence. i.e., any closed surface introduces a notion of mass because no continuous deformation can make such curvature go away: both a sphere and the surface of a cube are curved, in a smooth and non-smooth manner respectively, and neither can be homeomorphically transformed into a plane.

Black holes and wormholes thus introduce novel paths through the lattice of three-tori, as the extreme limit of Einstein's gravity theory, and their characteristic radius equates to a measure of mass. How they can be a natural element in the construction of quantum spacetime is discussed in the remainder of this monograph. In any case, black holes and wormholes are *non-local* expressions of inertia, and they are the quantum spacetime equivalents of Mach's principle.

The main result to keep in mind is that general relativity has an implicit quantum aspect through the entanglement of observer and observed, i.e., an operator who mimics. This drives the need to share quantum information globally along all paths/histories. A notion that the lattice of three-tori can provide for. In this, special and general relativity are preserved and the lattice of three-tori extends them with a multiply connected spacetime topology of distinct yet equivalent quantum particle trajectories. All the above points are returned to in the remainder of this monograph. First, motivated by Mach's principle, it is time to explore black holes and wormholes as massive expressions of quantum spacetime.

# Chapter 5

# Properties of Black Holes and Wormholes

As gravity bends spacetime, one can conceive of the possibility that the four-space continuum is curved beyond the point that an observer is still able to find a path that releases him/her from a gravitational potential well. Consider the point mass $M$ from before. In a simple Newtonian approach one may equate the kinetic energy of the observer with mass $m$, so $\frac{1}{2}mv^2$ to the gravitational work done from a radius $r$ to infinity, $GMm/r$. One has that $\frac{1}{2}mv^2 = GMm/r$ and therefore $r(v) = 2GM/v^2$. The ultimate speed of escape is that of light and $v = c$ yields $R_S = 2GM/c^2$ for the characteristic radius until which one can approach the point mass and still escape. Note in this that one deals with a point mass, which assumes of course that the observer is always small enough to fit comfortably within a region of radius $R_S = 2GM/c^2$.

Using general relativity, one can consider paths through spacetime and continue the discussion on time dilation in the previous chapter. Recall the temporal metric component $g_{00} = 1 - 2GM/(rc^2) = 1 - R_S/r$, for the characteristic radius $R_S$ encountered above. As pointed out before, $g_{00}$ becomes zero for $r = R_S$. In this, $R_S$ is the so-called Schwarzschild radius.[1] This is precisely the same result as a simple Newtonian analysis, but much more can be learned from it, particularly when again appreciating the importance of paths.

[1] Named after the person who thought of it while fighting in the trenches of the first world war.

# 1. Black Holes

First, consider an observer who is stationary near $r = R_S$. The limit $g_{00} \to 0$ signifies that for some finite proper time $\tau$, the coordinate time $t$ of some distant observer goes to infinity, i.e., $d\tau^2/(1 - R_S/r) = dt^2$. Hence, according to the distant observer, time comes to an almost complete stop for the loitering observer when the latter sits in a gravitational potential well under the condition that $r \to R_S$. For example, any actions that the stationary observer performs near $r = R_S$ go in extreme slow motion as he/she is observed from far away. Hence, any signal that is sent to a distant observer, including the stationary observer himself/herself embarking on a return trip, has an ever decreasing chance to escape from close to a sphere $S^2$ of radius $R_S$.

Consequently, this $S^2$ surface is called the *event horizon* and no causal communication is feasible, within the traditional framework of general relativity, beyond this boundary. Such an object, the ultimate expression of spacetime curvature, is called a black hole. If one computes $R_S$ for the mass of the Earth, then one finds about 1 cm. This while $R_S$ is roughly 3 km for the Sun. Obviously, the Earth and Sun need to compressed a lot for them to turn into black holes and there is little risk to us.

Second, for an in-falling observer who moves along a radial path to the event horizon, one should also consider the stretching of space. Indeed, the metric $g_{\mu\nu}$ contains elements like $g_{11}$, $g_{22}$ and $g_{33}$. In this, one may use the radial symmetry of a non-rotating black hole and take all off-diagonal metric coefficients as zero (e.g., $g_{01} = 0, ..., g_{23} = 0$). Thinking back to the rocket thought experiment, bending of space is experienced for observers moving along paths that take them to the vicinity of the black hole.

In this, it is convenient to use spherical coordinates $t$, $r$, $\theta$ and $\phi$ for the respective indices 0, 1, 2, and 3. The latter two are then angles that indicate two directions on the surface $S^2$. They thus navigate as longitude and latitude on the surface of the Earth. Hence, for these angles one can simply write a distance contribution as $r^2(d\theta^2 + sin^2\theta d\phi^2)$, so $g_{22} = r^2$ and $g_{33} = r^2 sin^2\theta$.

The radial metric coefficient $g_{11}$ can be determined straightforwardly, when one appreciates the non-essential nature of coordinates in general relativity. A stationary observer near the black hole is characterized by $dr = d\theta = d\phi = 0$. Conversely, an in-falling observer who measures his/her radial path length with respect to a single fiducial moment in time obeys $dt = d\theta = d\phi = 0$. That is, path length $r$ can be chosen by the in-falling observer to track his/her route. Consequently, the coordinate time interval $dt \rightarrow 0$ because any change in path length refers to some fixed temporal zero point, say related to the distant observer unaffected by the black hole. One can even think of a flexible rope that the in-falling observer ties to himself/herself and fixes to the distant observer, before plunging in.

So the paths of the stationary and in-falling observer respectively *choose* $t$ and $r$ to parameterize a specific aspect of their history through four-space, being duration versus length. Of course, the stationary observer is only at rest with respect to the event horizon and time still passes for the in-falling observer. These are merely coordinates. Inevitably, there must be a unique relation between these two routes. As long as the black hole is not rotating, the two choices are complementary, as can be seen by the consideration of light signals.

Suppose that both the stationary and the in-falling observer carries a standard, prior calibrated, light source to exchange information with the distant observer. A stretch in time $g_{00}$ of the stationary observer expresses itself in a lower frequency and thus a longer wavelength detected by a distant observer. A stretch in space $g_{11}$ of the in-falling observer results in a longer wavelength and thus a lower frequency detected by the same distant observer. This is a symmetric situation.

Then consider the situation where the in-falling observer takes a path from the distant observer at location A to the stationary observer near the horizon at location B. Upon arrival at the event horizon, the in-falling observer cannot be distinguished from the stationary observer by the light signals that either emits. This while the stationary observer only has measurements in $dt$ and the in-falling observer in $dr$. Hence, the change in proper path length, expressed through $g_{11}$ and experienced by the in-falling observer from A to B, must be the *inverse* of the change in proper time, expressed through $g_{00}$ and carried by the photon signal from B to A. Consequently, $g_{00}g_{11} = 1$ and $g_{11} = (1 - R_S/r)^{-1}$. In all, it follows that the so-called Schwarzschild metric of a non-rotating black hole obeys

$$d\tau^2 = (1 - R_S/r)dt^2 - (1 - R_S/r)^{-1}dr^2/c^2 - r^2/c^2(d\theta^2 + sin^2\theta d\phi^2). \quad (5.1)$$

One sees from the above expression that there are two forms of singular behavior. The $g_{00}$ coefficient blows up for $r$ near zero and the $g_{11}$ coefficient does so for $r$ around $R_S$. It is easy to see how to interpret these limits 1) $r \to R_S$ and 2) $r \to 0$. Working from the outside in, the former limit is considered first.

1) A distant observer cannot actually see a traveler fall into the black hole. This is due to the fact that the wavelength of any electro-magnetic signal that the in-falling observer emits, is gravitationally red-shifted as well. That is, as the clock of the in-falling observer appears to slow down for a distant observer, so does the frequency of any light source carried by the in-falling observer become lower due to gravitational redshift. In fact, for a free-falling body there is both gravitational and kinematic time dilation, so two factors of $(1 - R_S/r)^{1/2}$. This is because $GM/r = \frac{1}{2}v^2$ sets the kinetic energy per unit mass of the in-falling body in the expression of the inverse gamma factor $\gamma^{-1} = (1 - v^2/c^2)^{1/2} = (1 - 2GM/(rc^2))^{1/2}$.

Of course, the in-falling observer himself/herself is oblivious to such temporal effects and he/she crosses the horizon all the same, so in a finite proper time. The in-falling observer does note that the distant observer's clock runs much faster than his/hers. Indeed, from the in-falling observer's perspective it is the rest of the universe that is evolving at a faster pace. Again, observed time is relative. An interesting and quite real effect then occurs when two observers synchronize their clocks and one makes a round-trip that takes him/her very close to the event horizon. Upon return of orbiting the black hole for a day from the perspective of the observer who stayed behind, the two observers can compare their clocks. Indeed, the clock of the traveling observer does not indicate a time lapse of a day. Rather, not more than a few hours have elapsed if his/her trip was close to the event horizon.

Furthermore, an accompanying stretch of space through the Schwarschild metric also applies to the in-falling observer, as watched by the distant observer, and the former thus appears to be pulled apart like spaghetti. Indeed, the in-falling observer is only able to find a locally flat region if the extent of his/her own body is small enough compared to the length scale on which the curvature pulls matter apart. So on his/her way to the $r = 0$ singularity, the in-falling observer is inevitably pulled apart radially at some point. The poor in-falling observer is also squeezed sideways to fit in the small volume around $r = 0$. This is an unpleasant experience.

Nevertheless, the in-falling observer and distant observer each have their own reference frame, and neither has any absolute standing. As such, the singular behavior of $g_{11}$ at $r = R_S$ is not physical to the in-falling observer. That is, the in-falling observer's path to $R_S$ and the distant observer's path at a stationary location $r >> R_S$ are distinct and employ different coordinate systems. As the name event horizon suggests, the proper experience of the in-falling observer is to cross the $S^2$ surface in his/her own time, leaving a fading image behind.

Also, as one approaches the event horizon the earlier approximation that $\Delta\Phi/c^2 = GM/(rc^2) << 1$ no longer holds. This does not diminish the above arguments, but emphasizes again that care should be taken whenever some coordinate (like $r$) is adopted. A full-fledged derivation using Einstein's equations finds exactly the same Schwarzschild metric. This is not so strange because in general relativity one actually needs a so-called asymptotically flat spacetime region, equivalent to a linear approximation $\Delta\Phi/c^2 << 1$, with which to properly define the black hole mass in the first place. i.e., the mass of a black hole only has meaning with respect to the fiducial masses of distant observers. Indeed, this is a subtle expression of Mach's principle.

2) There is also a real singularity, though. As the in-falling observer reaches $r = 0$, in principle undetectable by the distant observer, $g_{00} \rightarrow -\infty$ and $g_{11} \rightarrow 0$. Hence, the proper time interval becomes unbound for any non-zero coordinate time interval. Of course, given the fundamental measurement uncertainty of our universe, one can never state that $r = 0$ is reached exactly. Only $r \approx l_P$ has significance in our universe. That is, the singularity hiding behind the event horizon should only be contemplated from a quantum perspective. This is the right mindset for what follows immediately below.

## 2. Leaky Event Horizons: Black Hole Temperature

So far, quantum effects have not been considered for black holes. Of course, one should consider those as well. Recall first the quantum uncertainty of matter and light. The location and momentum of all particles is undetermined to some degree, through the Heisenberg uncertainty principle. This is an expression of quantum spacetime because it consists of many different paths that every particle follows simultaneously when it travels from four-space point A to four-space point B. In this manner, matter and light also behave as waves because paths are separated spatially and temporally. The resulting effect is quantum superposition, so being everywhere in a fleeting sense.

Although the probability might be small, this also means that particles can move as waves through the black hole event horizon. The physical reason is again the lattice of three-tori. Its multiply connected nature allows for paths through the fourth dimension of time. This way matter and energy absorbed by the black hole can be re-emitted at a later moment. In effect, there is an *escape* from a black hole, thanks to quantum spacetime topology.

Any matter or energy that leaves the black hole diminishes the mass of the hole. Hence, quantum effects allow a black hole to evaporate mass that it has accumulated in its past. An effect first suggested by Hawking. It is not obvious in what form matter or energy exits the black hole. If an entire observer comes out in one piece, then all the better. However, in practice one expects that macroscopic objects come out in bits and pieces. Still, for the notion that information in some form can actually traverse a black hole this is not relevant. A way out is a way out and black holes have *leaky* event horizons.

As do all radiation effects, this emission process has a characteristic temperature $T_{\rm BH}$ as well. After all, if a black hole radiates, then it is not completely black with a temperature of zero. This is a direct consequence of thermodynamics. Anything that produces some form of heat must have a non-zero temperature. In a purely general relativistic sense it is not easy to compute this temperature. The quantity that one seeks, is the surface gravity. So the acceleration very close to the event horizon, since that creates the energy (i.e., the heat) for emission in the rocket thought experiment. Indeed, the units of acceleration are length divided by time squared, which can be converted into an energy by multiplication with $h/c$, or a temperature through $h/(ck_{\rm B})$ for $k_{\rm B}$ the constant of Boltzmann. Of course, gravitational red-shift, as discussed above, strongly suppresses any such observational signature for a distant observer. The balance between the two is therefore not obvious.

However, in a topological and quantum sense one merely needs to adopt the perspective of the rocket operator. i.e., recall that any observer-observed entanglement, necessary for an actual quantum observation, proceeds through the comparison of distinct yet equivalent paths. To establish the emission properties of a black hole one can use, for the first time in this monograph, that important principle. Because of it, there must be two distinct yet equivalent paths through which the black hole acquires a thermodynamic temperature as perceived by a distant observer.

To this effect, consider an operator who mimics the acceleration at the event horizon and *also* put him/her in the location of a distant observer, as follows. The operator is already the observed in Einstein's elevator thought experiment. This is because it is his/her actions that are experienced. Nevertheless, one can also allow the operator to witness the results of these actions himself/herself at a safe distance from the event horizon. After all, given two observers, either can play the operator or the distant observer. The two resulting four-space histories are distinct yet equivalent, whatever paths are chosen by the observers. Hence, one can freely interchange the roles of the operator and the distant observer, so within either history, without violating tangible reality. The perspective that *observer and observed mimic each other*, is the heart of quantum mechanics and also the key to black hole thermodynamics!

In this, the operator, when acting as the distant observer, mimics the acceleration effect under Newtonian conditions because he/she is far away from the black hole event horizon, in an asymptotically flat spacetime region with $r >> R_S$. The radial acceleration that the operator needs to produce is then $g = GM/r^2$. Because the event horizon perspective is mimicked, this implies that $r = R_S(M)$ and thus $g(M) = GM/R_S^2$. Consequently, the surface gravity $g = 1/(4M)$ for a Schwarzschild black hole of radius $R_S = 2M$, when one writes all quantities in so-called natural units with $\frac{h}{2\pi} = c = G = k_B = 1$.

The key point here is that the *availability* of the information that $r = R_S = 2M$ is a direct consequence of the non-local connections in the lattice of three-tori. i.e., the two four-space histories are in global superposition, with the properties of the black hole accessible to the observed-operator-acting-as-a-distant-observer. In this, the exchange of quantum entanglement information proceeds through an $S^1$ loop, so a closed four-space path in the lattice of three-tori, as follows.

The distant observer is also a mimicking operator and needs information on the event horizon. This requires a round-trip to exchange the information. Also, the paths for the distant observer and the observed operator must be *locally* the same by the very construction of the thought experiment. After all, one has merely assigned two different labels to the same local entity. Nevertheless, this entity needs to possess two different identities, one as observer and one as observed, simultaneously. Therefore, the two distinct yet equivalent four-space paths that this requires, can only exist *globally*. These must thus be the result of a topological identification through time, with the two four-space paths forming a single $S^1$ loop.

Consequently, the surface gravity responsible for the black hole temperature is spread across the circumference of an $S^1$ loop. Given that $k_B T_{BH}$ equates to an energy, multiply connected quantum spacetime literally diminishes the heat that results from the black hole's surface gravity because observers and observed can be exchanged by an $S^1$ loop through the fourth dimension of time. Hence, $T_{BH} = g/(2\pi) = 1/(8\pi M)$, with a proper $2\pi$ normalization due to such a loop. Besides the important distinction between $N$ and $N'$ when it comes to the spatial characterization of particles (in integer units), one therefore also finds that multiply connected time renders tangible reality truly non-local in $2\pi$ cycles.

One further has that $T_{BH} = 1/(8\pi M)\ \frac{h}{2\pi}c^3/(Gk_B)$ with all constants included explicitly, rather than using natural units. So the thermodynamic temperature of a black is inversely proportional to its mass. The more massive the black hole, the colder it is. This is fascinating but also quite plausible thermodynamically. More mass leads to a proportionally larger Schwarzschild radius. Such a larger length scale, when viewed in a photon sense, in turn implies a longer wavelength. As is well known for a blackbody radiation field, its peak wavelength $L$ scales as the inverse of its temperature. So $T_{BH} \propto L^{-1}$ on thermodynamic grounds. This recovers the result above for $L \to R_S$ and indicates that black holes behave as blackbody emitters. Of course, it is the $2\pi$ normalization that expresses the beauty of observer-observed mimicry.

If one would have done this computation in a purely classical sense then the result above for $T_{BH}$ would not have had any merit because Newtonian mechanics has no validity near the event horizon of a black hole. Ironically, the inclusion of quantum effects through the entanglement of observer and observed, together with general relativity through the equivalence principle and an operator, indicates that the naive classical surface gravity result is actually correct. The important point to remember in this is that observer-observed mimicry through distinct yet equivalent quantum spacetime paths has deep physical meaning. This point is returned to in subsequent chapters.

## 3. Wormholes

The mere fact that a black hole is leaky, through multiply connected quantum spacetime, allows it to behave as a wormhole. The latter is also called a handle and is denoted by $S^1 \times S^2$. The surface $S^2$ is again the event horizon, while the loop $S^1$ now indicates that a wormhole constitutes a *tunnel*. i.e., the mathematical form $S^1 \times S^2$ highlights the most important property of a wormhole/handle: it has a throat to pass through from a four-space point A to a four-space point B. In this, points A and B can be far apart in spatial location and/or time.

As mentioned before, the lattice of three-tori augments the $S^2$ event horizon of a black hole with paths that move back and forth through its surface. The starting point A and end point B of any four-space path need not be the same in space and/or time. In fact, paths entering the $S^2$ surface and those exiting it trace two distinct spherical surfaces, one entrance and one exit per wormhole. Such a wormhole is thus a three-dimensional object that requires four dimensions to be embedded, so that it can exist as a spacetime construct. A wormhole is sometimes called an Einstein-Rosen bridge and one may simply view it as two black holes that are connected to each other through the fourth dimension.

Both a non-evaporating black hole and a wormhole are bona fide solutions of Einstein's general relativity equations. The non-evaporating black hole is an immediate result, as shown already by Schwarzschild and the corresponding metric above. Within Einstein gravity, and ignoring quantum effects, one can view the paths of in-falling observers equally well as out-going ones. This because the $r = R_S$ singularity is not physical, so one may simply exchange the inside and outside of the event horizon. This then yields a wormhole solution. The wormhole is also a natural expression of geometry plus quantum uncertainty on the Planck scale. i.e., for a mass $m_P$ one has that $R_S(m_P) = l_P$. This is explored further in chapter 7 (in the form of Wheeler's quantum foam for the already curious reader).

Furthermore, locally a non-evaporating black hole and a wormhole look the same in terms of geometry. However, a wormhole can topologically identify different regions and/or different times. It is thus an intrinsically much more non-local entity. A wormhole topology (i.e., a tunnel through four-space) is different from a non-evaporating black hole topology (i.e., a surface enclosing a singularity of no-return). In any case, with a handle there are two topologically distinct paths. One can go through the wormhole from A to B while the other one goes around the wormhole from A to B and does not cross any event horizon. These two paths cannot be obtained from one another through continuous deformation, i.e., they are globally different while locally the same.

So a wormhole is topologically distinct from a non-evaporating black hole. However, a black hole is actually leaky and evaporates, as argued above. That is fundamentally different from a non-quantum black hole. In fact, a wormhole transfers mass and/or energy to a different place and time as something traverses its mouth. This while an evaporating black hole that moves with respect to some observer absorbs matter at one place and time and re-emits it, in some form, at another place and time. As such, a leaky black hole behaves as a wormhole.

Also, the topology of a wormhole and a leaky black hole are precisely the same: a path goes in at spacetime point A and comes out at spacetime point B by passing through the fourth dimension. Hence, for the remainder of this monograph, the term black hole should be read as leaky black hole or quantum black hole, indicating that all black holes act as wormholes through the lattice of three-tori. Sometimes the terms leaky and quantum are mentioned explicitly to aid the reader.

The nature of the emissions along the leaky black hole's four-space path is pertinent of course. A completely incoherent flow of particles, even though a perfectly reasonable history in spacetime, is a far cry from what popular fiction likes to perceive as a wormhole.[2] Nevertheless, in terms of matter traveling through spacetime, a leaky black hole of mass $M$ fits all the criteria of an $S^1 \times S^2$ handle topology embedded in four dimensions. In addition to this, both geometry and topology play a role, for the following reasons.

A macroscopic wormhole requires a large value of $R_S$ and thus a large mass. In a purely general relativistic sense, this means that creating a larger wormhole inevitably enhances its lingering identity as a non-quantum black hole. This is because an increase in mass diminishes the black hole temperature proportionally and thus its ability to evaporate, i.e., to sustain a four-space connection. In this, the singularity at $r = 0$ of the Schwarzschild solution suggests that anything larger than $l_P$ and more massive than $m_P$, the natural scales of the four-space passage, would have a difficult time coming through, as follows.

[2]Anything that emerges from the "other side" of a black hole, in the form of evaporation or through a wormhole bridge, is also referred to as a white hole.

A mere kilogram of matter already amounts to about $10^5 m_P$ and is therefore a large gravitational disturbance to the Planck scale singularity. The total mass of the black hole, say for a size of 10 m, is about 1,000 times the mass of the Earth. Hence, as matter falls in for $r < R_S$, an observer traveling with it has very little chance of making it through the wormhole as the proper time blows up for $r \to 0$. Any traversing matter thus seriously affects a macroscopic wormhole configuration by de-stabilizing it, literally pinching it off, for time scales much longer than $t_P = l_P/c$.

The above indicates that one should explore the nature of wormholes as passages not just through spacetime curvature (so geometry) that becomes singular, but through the lattice of three-tori (so topology) that allows for a Planck scale description. After all, the lattice of three-tori leads to the quantum properties of black holes in the first place. Hence, to create a bona fide wormhole one needs to contemplate the global aspects of the lattice of three-tori and how it *transforms* non-quantum black holes into wormholes (so leaky black holes) using topological identifications (so portals through time that connect different histories/realities). Indeed, the topology of quantum spacetime should be formulated in a dynamical sense, with full consideration of an evolving four-space connectivity. These issues are returned to in the following chapters.

## 4. Mach Again

Finally, it is useful to discuss some more on non-quantum black holes and wormholes in the context of Mach's principle. Non-quantum black holes have a clear inside and outside in a general relativistic sense. Their properties are therefore quite similar to those of particles in that they have a mass, an electric charge and a spin (so rotation). Because of these limited and basic properties non-quantum black holes are like particles with a size $R_S$. Despite this extended nature they are well localized in space and time because they lack true quantum four-space properties when one ignores evaporation through the lattice of three-tori and observer-observed entanglement.

Wormholes, so leaky black holes, on the other hand are truly extended massive objects. That is, they carry a notion of inertia in a global sense because the four-space points A and B that they connect can be far apart for an observer. Furthermore, a handle is a three-dimensional $S^1 \times S^2$ manifold embedded in four dimensions. A universe that comprises many such wormholes can therefore be viewed as a global four-space network that induces mass locally through its $S^2$ surfaces and $S^1$ loops.

Consequently, whatever an observer perceives as inertia locally, in a quantum spacetime it must be the expression of the wormhole collective. This is because observer mass and observed mass are unremittingly entangled as an expression of mimicry. Hence, mass is purely topological in Mach's sense. It is carried locally through the presence of $S^2$ spherical enclosures and is generated globally through the existence of $S^1$ loop connections in four-space. This approach to the concept of inertia appears close to the spirit of what Mach alluded to. In the remainder of this monograph it is shown how matter in the form of quantum black holes can generate observable gravitational effects beyond Einstein gravity, like the currently popular dark energy that drives the expansion of our present-day universe. First, it is time to explore some of the properties that our universe possesses.

# Chapter 6

# Cosmic Evolution: A Black Hole Perspective

It is important to have an overview of the kind of world one lives in. After all, it is the evolution of the cosmos that has inspired so many to ponder and understand it. In particular, the nature of our universe allows for the formation of black holes, be they very large or very small. When dealing with quantum spacetime, and as discussed above already, leaky black holes take a special place. In this, gravity is the most important force that shapes the evolution of our universe. Thanks to gravity there are galaxies, stars and planets. A testimony to gravity and dynamical spacetime is the expansion of our universe. In every direction one sees, on scales of millions of light years, galaxies recede at a speed that is roughly proportional to distance.

The cosmos is about 14 billion years old and as one goes back in time there is the inevitable big bang, or at least an edge to our perception of time. Our universe came into existence by this big bang, a singularity that unraveled itself through a phase in which the four fundamental forces (gravity, electromagnetism, weak and strong nuclear interaction) were unified, a so-called theory of everything. The presence of gravity, an expression of spacetime as explained before, signifies a very important aspect. With the big bang not only matter became pertinent, but also the space and time that it occupies. In other words, one cannot ask what there is outside or before our cosmos. Spacetime itself, with matter as its occupant, emerged with the big bang.

Very shortly after the inception of our universe, the four fundamental forces were likely of comparable strength. This is the so-called quantum gravity or Planck era less than $10^{-43}$ s (so the planck time $t_P$) after the big bang. After this Planck phase a period of rapid expansion of the cosmos, called inflation, followed during which our universe grew exponentially in size by a factor of about $e^{55} \sim 10^{24}$. Our universe was then $\sim 10^{-36}$ s old. Through adiabatic expansion, so without heat exchange with an exterior environment (befitting a closed system), the universe cooled down. Below about $10^{16}$ GeV[1] ($\sim 10^{-3}$ times the Planck energy) grand unified theory, the unison of the three non-gravitational forces, was broken in a cosmic phase transition.[2] Much later in time, but still within a millisecond, there were quarks (the building blocks of protons and neutrons), leptons (e.g., electrons) and an abundance of photons as a consequence of a small imbalance in matter and anti-matter.

Elements like hydrogen and helium were formed through nucleosynthesis (protons and neutrons) in the first three minutes. During the 400,000 years after this, atomic nuclei, electrons and radiation were in equilibrium with each other in the form of a plasma (an ionized gas). It was in this period that the cosmic microwave background, which we can observe until this day, was formed. The current temperature of this blackbody radiation field is about 2.7 K. From 400,000 to 100 million years after the big bang the universe comprised mostly neutral hydrogen, without any stars. This period is aptly called the dark ages. The first stars were formed after about 100 million years and the cosmos has subsequently been busy for billions of years to form galaxies. Hence, the stars that one sees today in the night sky have not always been there.

Furthermore, there are roughly $\sim 10^{19}$ black holes in the observable universe, with masses of $1 - 10$ solar masses and called stellar origin black holes. Besides these there can be small quantum black holes with masses as modest as $10^{-5}$ g (the Planck mass) to more than $5 \times 10^{14}$ g. Recall that the terms wormhole, leaky black hole, handle and quantum black hole all indicate the same (topological) object: $S^1 \times S^2$. This while the term black hole always implies quantum (so wormhole) properties through $T_{\rm BH} > 0$ K. In this, the Planck mass

[1]One GeV is approximately equivalent to the mass of a proton or about $1.6 \times 10^{-24}$ g times $c^2$.

[2]In fact, at approximately 246 GeV (roughly 200 times the mass of a proton) the weak force and electromagnetism split off from the electroweak force.

is the mass that yields a Schwarzschild radius equal to the Planck length, a mini black hole indeed. Also, almost every galaxy contains a supermassive black hole in its center, ranging from $10^5$ to as much as $10^{10}$ solar masses. These monsters have been detected less than 900 million years after the big bang and produce large amounts of radiation through the accretion of matter. In all, the most abundant manifestations of gravity are stars and black holes, with $10^{22}$ stars distributed among $10^{11}$ galaxies in our observable cosmos.

## 1. Gravitational Collapse: Different Outcomes

A gravitating system is stable if the gravitational pressure exerted by matter is compensated by the kinetic energy of material particles and/or the momentum of photons. This suggests an instability when the internal pressure of an object is insufficient to carry its own weight. A criterion for instability is the so-called Jeans mass, $M_{\mathrm{J}} = (T/10)^{3/2}(\rho/10^{-20})^{-1/2}\ M_{\odot}$ in solar masses $M_{\odot}$ (about $2 \times 10^{33}$ g), for gas temperature T in K en mass density $\rho$ in g cm$^{-3}$. That is, the Jeans mass sets the characteristic scale on which gravity takes over and collapse ensues.

An almost homogeneous medium with small perturbations (like the ones introduced right after inflation, see chapter 9) therefore collapses into fragments with mass $M_{\mathrm{J}}$. During contraction the density increases. So at a constant temperature, $M_{\mathrm{J}}$ decreases further and further during collapse and objects of much less than a solar mass can be formed. However, the temperature increases if the collapsing gas cannot radiate away the gravitational potential energy that is released by contraction and the deepening of the potential well. In this fashion a star is formed that can radiate through nuclear fusion for many millions of years. A gas cloud with a density of $10^{-19}$ g cm$^{-3}$, a typical initial state 100-500 million years after the big bang, takes about $10^5$ years to collapse into a star.

Collapse can also be singular. i.e., a black hole forms instead of a star. These black holes are called monolithic. They result when the collapse proceeds more or less adiabatic, so without the exchange of heat (radiation) with the environment. The Jeans mass then remains constant even though the density increases. There is little fragmentation, while $M_{\mathrm{J}}$ in the early universe can be as high $10^3 - 10^5\ M_{\odot}$ because the gas temperature is more than $10^4$ K. So much mass cannot be carried by the hot core of a star. The gas cloud subsequently collapses until spacetime is curved onto itself by the gravitational field, following Einstein's theory of general relativity, and a black hole is formed as discussed above. The radius of the event horizon is $R_S = 3(M/M_\odot)$ km, for the mass $M$ of a black hole in solar masses. This simply is the Schwarzschild radius written in more astronomical terms.

To understand the origin of black holes today, and thus their role in quantum spacetime, it is necessary to consider very early times as well. In the first epochs of the universe, less than $10^{-36}$ s after the big bang (the Planck and inflation phases), black holes are also formed. These small black holes are initially quite light, less than $10^{-3}$ g, because the density is huge right after the big bang. This lowers the Jeans mass more than the temperature makes it rise, even though the temperature is about $10^{31}$ K. Such primordial quantum black holes can grow to more than $10^{15}$ g in the first hours after the big bang. Hence, black holes in our universe can be very small and very large in terms of their Schwarzschild radii $R_S$.

## 2. Mini versus Maxi Black Holes

It is important to consider the nature of mini (Planck scale) and maxi (stellar and monolithic) black holes, and any mass in between. As derived above through the lattice of three-tori, every black hole has a temperature $T_{\rm BH} = 10^{-7}(M_\odot/M)$ K, with the expression from above written again in more astronomical terms. The event horizon is thus leaky and black holes can even evaporate completely through the emission of this so-called Hawking radiation. A black hole then exists for at least $t_e \sim 10^{56}(M/M_\odot)^3$ billion years.

This is easy to understand when one realizes that a black hole comprises many particles and thus constitutes many degrees of freedom. Under such circumstances, the radiation process should reflect that of an object in thermal equilibrium, radiating as a black body, as follows. According to the law of Stefan-Boltzmann, the power emitted by a black body scales as its surface area $A$ times temperature $T_{\rm BH}$ to the fourth power. For black holes $A \propto M^2$ and $T_{\rm BH} \sim M^{-1}$. Because the emitted power must expresses the mass loss rate of the black hole, $dM/dt \propto M^{-2}$ and $t_e \propto M^3$. A black hole of more than $\sim 5 \times 10^{14}$ g therefore lives for about the current age of the universe, 14 billion years, and could be evaporating during our epoch. In this, one should realize that the omnipresent cosmic microwave background of $\geq 2.7$ K typically has a temperature that is higher than $T_{\rm BH}$ for $M \geq 10^{-7} M_{\odot}$, suppressing evaporation of stellar origin and monolithic collapse black holes on thermodynamic grounds.

Massive black holes are also very much capable of attracting matter and thus to grow in mass and lower their blackbody temperature. This accreted matter spirals to the black hole and the associated friction heats the gas to more than $10^6$ K. At this temperature matter produces copious amounts of ultraviolet and X-ray radiation before it is absorbed by the event horizon. Therefore, it is friction heat of accreting material that makes massive black holes the power houses of our universe, while they sit in the nuclei of galaxies.

In this accretion process, less than 50% of all absorbed mass $m$ can be converted into energy $E$ through Einstein's famous formula $E = mc^2$ (recall that $c = 300,000$ km s$^{-1}$). This is a theoretical maximum that follows simply from consideration of the gravitational work $W$ performed by a black hole with mass $M$ as matter comes from very far (approaching infinity) and goes into orbit at a radius $r$, so $W = GMm/r$. When one divides the latter by $mc^2$, then this is the radiative efficiency $\epsilon = W/mc^2$ with which mass $m$ is converted into energy $E$ when the gravitational work performed on in-falling matter is completely dissipated as radiation at a surface with radius $r$.

This yields $\epsilon = R_S/2r = \frac{1}{2}$ for $r$ equal to the Schwarzschild radius $R_S = 2GM/c^2$, which defines the most compact object within general relativity. A non-rotating black hole typically gets to not more than 16%, and closer to 6%. This is because the deepest stable orbit for matter lies at $r = 3R_S$. Within this orbit material that moves toward a smaller radius must spiral in further. A very fast rotating black hole can reach $\epsilon$ values of a bit more than 30%. In all cases this is much more than the $\sim$ 1% that our Sun achieves through the nuclear fusion of hydrogen to helium.

When black holes meet each other they can merge into a black hole whose mass is approximately the same as the sum of the original black holes. However, a part of the mass can be converted into gravitational waves, i.e., ripples in spacetime that are created by the spiralling-in motion of the binary black hole. These gravitational waves travel at the speed of light and astronomers are trying to detect these as another confirmation of Einstein gravity with detectors on Earth. Many small ones thus also make a large one. Through merging, a collection of small black holes can grow into larger black holes, suppressing their leaky (so wormhole) nature. This while larger black holes can grow into supermassive ones when aided by accretion. In any case, Planck scale, primordial, stellar and monolithic black holes all have their own origins and histories. Still, it is together that they define the global topology of spacetime in terms of $S^1 \times S^2$ handles.

Black holes have special meaning through Mach's principle as highlighted before. As such, it is important to know how abundamt they are in our cosmos. There are about $10^{19}$ black holes in the observable universe, of stellar origin, given the currently very large number of stars ($\sim 10^{22}$), as follows. Stars in our own Milky Way are typically 0.5 $M_\odot$ and live for 10 billion years or more. However, about 1% of all stars are more massive than 8 solar masses. These massive stars explode as so-called type II supernovae after only $10 - 100$ million years. They leave a remnant behind that is most often a neutron star, but sometimes a black hole, with a 10% probability. For our Milky Way galaxy, the overall probability is therefore only 1 in $10^3$ to end up with a black hole if one picks a star at random. Nevertheless, with $10^{22}$ stars in the observable universe, this still gives a very large number of $10^{19}$ black holes.

The present number of black holes in the universe can be even larger if numerous primordial black holes are created during the Planck and inflation phases of the universe, less than $10^{-36}$ s after the big bang. Such primordial black holes can survive until the present day only if they are more massive than $\sim 5 \times 10^{14}$ g, as discussed above. However, observational limits on such evaporation events suggest that stellar origin black holes dominate by number in our universe.

In all, there are many origins for black holes that are long-lived under quantum evaporation, with their creation intimately tied to cosmic evolution. Because the bulk of all black holes is expected to be stable, a priori wormhole characteristics are likely to be modest because of tiny thermodynamic temperatures. However, as already mentioned in the previous chapter, it is essential to also consider the *dynamics* of quantum spacetime connectivity in terms of quantum black holes and three-tori. This is the final step to make in understanding the true nature of quantum spacetime and the next chapter describes it.

# Chapter 7

# The Mimicking Nature of Quantum Spacetime

With a proper understanding of our cosmos, it is time to describe the very heart of quantum spacetime. To this effect, recall that measurement uncertainty, a universal property, requires Nature to construct all sorts of topologically distinct paths between any two four-space points A and B. This yields a four-dimensional lattice of three-tori that experiences continuous deformations, which are called homeomorphisms. This spacetime also hosts a very large number of quantum mechanically leaky black holes, which consequently behave as wormholes.

Hence, Nature connects locations in space and moments in time through a lattice of three-tori and a collective of handles. The embedded $S^1$ loops, so the temporal obstructions (portals through time) that constitute inductors of entire histories for exchangeable observer-observed pairs, then globally distinguish paths without precise local measurements. All this allows leads to rather unexpected journeys of particles. Indeed, there is always a different path, befitting both Feynman's quantum path integral as well as Einstein's geometric gravity. These elements like three-tori and wormholes therefore allow for the most fundamental quantum aspect of the universe to be expressed: observer-observed entanglement.

Indeed, because any entity can take on the role of the observer or the observed, and both are required to jointly define an actual observation, observer and observed each follow a four-space path that is equivalent to the other locally yet is also distinct from the other globally. Just think back to the black hole temperature example and the mimicking operator's $S^1$ round-trip. Topological identifications through time allow for this idea to be implemented and thus for a guiding physical principle to be explicitly formulated. The latter is the task at hand now.

# 1. The Planck Scale and Topological Dynamics

In our universe, particles are the building blocks of matter. They are continuously created and destroyed. This while paths are the discrete fibers in the fabric of spacetime. Hence, it is the proper philosophy to consider the dynamics of paths, referred to as *topological dynamics* from here on, when it comes to quantum spacetime. From a physical perspective, the argument is equally straightforward. Quantum spacetime itself is also undetermined in its structure, i.e., in its topology. Consequently, spacetime undergoes quantum fluctuations in its connectivity. Indeed, the global histories that four-space paths constitute are unremittingly created and destroyed as part of the ever-changing spacetime fabric.

The starting point for topological dynamics is again the Planck scale. Imagine zooming in on spacetime with an apparatus all the way to the Planck length of $\sim 10^{-33}$ cm. Because of quantum uncertainty in the motions of particles and their associated curvature, one finds all sorts of geometric shapes. The latter befit the continuous deformations, so homeomorphisms, discussed above. A mathematical way of viewing this is that gravity, as expressed by general relativity, is not invariant under conformal transformations. These transformations change the scale of a system, e.g., when going from a meter to $l_P$. Hence, they are a type of homeomorphism that is isotropic. One merely re-scales all metric coefficients with a positive factor, keeping angles between vectors fixed. There are certainly other homeomorphisms to express quantum uncertainty on the Planck scale, but the conformal one follows directly from Einstein gravity. This is because curvature is proportional to mass density, a scalar quantity. Indeed, a local observer that sees space and time stretched or contracted conformally, concludes that a different gravitational acceleration field is at work because the density changes. Without going into the details of these conformal transformations, it is enough to appreciate the notion that quantum uncertainty in any measure of scale must fundamentally affect the geometric (so shape) dynamics of spacetime.

But there is more then. When quantum uncertainty is applied to the fabric of spacetime itself in this manner, then it must also undergo topological (so connectivity) fluctuations in four dimensions. After all, some of these geometric fluctuations may yield transient black holes. This while so far a fixed topology of spacetime has been adopted to express quantum mechanics, being a lattice of three-tori with leaky black holes. Hence, it is time to work toward a topologically changing four-space continuum in which paths (i.e., entire histories) are unremittingly created and destroyed. It is this dynamically changing connectivity, on the Planck scale as well as on larger scales, that is the true nature of our quantum spacetime. When constructed, it must breath the spirit of Mach's principle and has to render the distinction between local and global inertia obsolete.

A first step to this self-consistent quantum spacetime is the black hole in the form of a wormhole, but now as a quantum effect on the Planck scale. The natural measures to perceive quantum spacetime on the Planck scale are $l_P$, $t_P$ and $m_P$, already encountered above. The significance of these units of length, time and mass is that they are the only ones constructible from $h$, $c$ and $G$. An immediate observation to make is that $l_P$ and $t_P$ are small compared to everyday life experience, but that $m_P$ is quite well imaginable and much larger than the typical masses of the proton ($\sim 10^{-24}$ g) and electron ($\sim 10^{-27}$ g).

The relevance of the latter is that the Planck mass equates to a black hole with a Schwarzschild radius of about the Planck length, see the formula for $R_S$ earlier. Hence, $m_P$ may comprise many particles, but it is above all an expression of extreme gravity and inertia that goes well beyond individual particles. In other words, on the Planck scale spacetime acquires a discrete particle-like nature that expresses itself through non-local leaky black holes of about a Planck mass. These are called mini black holes or mini wormholes. Both terms are again used interchangeably.

In fact, on the Planck scale a mini black hole has such a high temperature that its evaporation time is about equal to the Planck time. As such, Planck mass wormholes are very transient phenomena. Indeed, the fundamental measurement uncertainties $l_P$ and $t_P$ are of the same order of magnitude as the mini black hole's size and life span. This is an important realization and one that leads to a very specific quantum spacetime property. Because of quantum measurement uncertainty, it is not possible to localize a mini black hole event horizon very accurately, nor to know when it has appeared or disappeared.

In effect, a mini wormhole is both present and absent on the Planck scale, as a result of measurement uncertainty. Even more pertinent, this special property of transient mini wormholes on the Planck scale pertains to every Planckian volume in the lattice of three-tori. Quite so, the two main constituents of Planckian quantum spacetime are now identified: the Planck size three-torus and the mini wormhole. Given its recent appearance in this monograph, it is prudent to first further explore the latter.

## 2. Wheeler's Quantum Foam and a Machian Perspective

The above collective of mini wormholes, as an expression of quantum uncertainty in Planckian spacetime itself, is called Wheeler's quantum foam (first perceived by him in 1957). These mini wormholes of mass $\sim 10^{-5}$ g evaporate in about $10^{-43}$ s, so in about a Planck time, because their $T_{\rm BH} \sim 10^{31}$ K. Since mini wormholes are such highly transient non-local entities, they appear virtual and not tangible to an observer. This while stellar and monolithic black holes acting as wormholes are many kilometers in size and are therefore almost eternal objects.

On the Planck scale, with so much uncertainty in the form of mini wormholes, a more or less stable quantum foam seems a tall order then. Indeed, a single Planck mass in a volume of $l_P^3$ represent a huge mass (or energy) density of $\rho_P \sim 10^{94}$ g cm$^{-3}$. Furthermore, these quantum fluctuations are an expression of spacetime itself, so of a vacuum with all matter removed. After all, in the arguments above matter degrees of freedom were not used. Quantum measurement uncertainty was simply applied to four-space itself, leading to a Planck mass in an Planckian volume for about a Planck time: a leaky mini black hole. Clearly then, the vacuum represents a potentially gigantic Planckian energy density of $\rho_P c^2$. This while macroscopic observers appear unaffected by it. So after the suggestion by Wheeler of the quantum foam, the immediate worry was how to perceive of this vacuum. Nevertheless, the arguments in favor of mini wormholes on the Planck scale cannot be dispelled either, being a logical expression of gravity and quantum uncertainty. Hence, one should embrace logic and follow the central thread of this monograph: *paths with mimicking observers and observed*.

Wheeler's quantum foam is what Mach would perceive as non-local inertia. In fact, since particles carrying mass and energy in effect travel along wormholes on the Planck scale, and there is no a priori preference of the paths in the lattice of three-tori over those carried by wormholes, particle inertia should actually be viewed as topological dynamics. Taking the plausible next step, one should seek to construct a quantum spacetime that allows for a world of connected three-tori, transient mini wormholes and leaky macroscopic black holes to carry all of tangible reality. Such a quantum spacetime should obtain its existence through a proper global natural law. i.e., through a law of mimicry rooted in paths such that their collective rules the individuals. This is what Mach, Einstein, Wheeler and Feynman would have liked (at least the author sincerely hopes so).

The inevitable perspective of mimicry is then that the *number* of paths that one perceives in some part of spacetime is the direct physical consequence of the total number of paths comprising the entire fabric of four-space. This must include creation and destruction of paths as dynamical spacetime entities. Hence, any observer in four-space is to conclude that all local changes in the number of paths, be they carried by three-tori (Feynman's path integral) or mini wormholes (Wheeler's quantum foam) or macroscopic black holes (cosmic evolution), result mathematically and physically from the global count of such paths. After all, in topology as the spacetime language of quantum mechanics, or through mimicry as a physical act of path multiplication, all quantities are expressed by integer numbers.

This topological formulation of Mach's principle is a lot to take in. It is also intentional by the author to expose the reader to this law of path dynamics at this point in the monograph. To appreciate a new perspective one needs time to get used to it, and one now has at least until the remainder of this monograph. There is one further step to make, though, before the central principle of quantum spacetime can be formulated properly and completely.

# 3. The Multiplication of Four-Space Paths

The key to a global construction of dynamical quantum spacetime lies in the entanglement between observer and observed. In the elevator thought experiment the operator needs to mimic a particular acceleration, for which specific information is required. In fact, the operator is a mimicker, so an active participant, and needs to navigate a path for the elevator to follow. Hence, whatever path the elevator follows, or a quantum particle in the double slit experiment, its own reality as an observed entity also amounts to an observable operator who mimics the very path that it follows.

In this, any observer can constitute an observed and every observed can be an observer. In fact, any particle or part of spacetime is both observer and observed, any time and any place. This argument holds *in general*, so not only within a gravitational context. After all, given a particular choice for an observed undergoing any acceleration or physical action, the observer may always choose to play the role of the observed and mimic the latter's situation by also undergoing the experience. Just think of Schrödinger's cat and imagine that the observer takes the place of the cat and that the latter observes the proceedings. That constitutes a globally distinct yet locally equivalent four-space history as far as natural laws are concerned.

Indeed, recall from above that an objective distinction between observer and observed, whatever the nature of their interaction, is impossible and that both activities are equivalent. This while any observer and observed, as an interacting system, must be entangled through a path that connects them as individual entities. Any such connection between two quantum particles or spacetime regions must therefore be realized with either particle or region acting as *both* the observer and the observed. Consequently, the only action that constitutes an observation in dynamical quantum spacetime is that of an observer mimicking an observed.

From above, this already led to the assertion that any observer-observed interaction can therefore only have quantum physical meaning if there are at least two topologically distinct connecting paths. This to realize the quantum equivalence of observer and observed, with one mimicking the other, while allowing both to have an individual identity with the ability to act on their own. The new light to view all this in, is that of a dynamical world where paths are unremittingly created and destroyed as the expression of mimicry. i.e., the identity of any path, as a history through four-space, always requires at least one other path for comparison as a joint observer-observed reality. This can be put in plain terms, as follows.

Imagine two children wanting to play cops and robbers. Either can be the cop or the robber and as such they are equivalent. Yet both activities require a distinct choice to be made, in favor of a cop or a robber, as the result of a child's personal history. Hence, the children's individual routes lead to a pertinent interaction because they are both distinct and equivalent. Therefore, it is the creation and destruction of these distinct yet equivalent children's four-space histories that yields a quantum world with entangled cops and robbers, as a joint expression of the children's individual identities.

i.e., when it comes to the identity of joint chains of events in spacetime, associated with an observer and observed, "it takes one to know one" so to say. Consequently, if paths are to be the individual building blocks of spacetime through their collective dynamics, then they need to be *multiplied*. i.e., the requirement of local observer-observed mimicry to establish the individual identity of four-space paths (so entire histories) in a comparative fashion can only be realized globally and dynamically through the multiplication of paths, so as children of children of children etc.

To tie all this together, consider the following thought experiment for an observer who wishes to assess the structure of his/her spacetime through some measurement process. The line of reasoning presented below is quite similar to the ones above leading to paths as the dynamical entities building up quantum spacetime, with small shifts in emphasis to work toward a fundamental physical principle and an accessible mathematical formulation. Excluding the philosophically interesting case of a completely isolated observer, it seems reasonable to assume that the observer can detect phenomena and come to some quantitative view of his/her universe through a measuring apparatus.

What properties of spacetime should the observer measure and how can he/she quantify these? Going back to the introduction of this monograph, it is good to recall the central tenet of quantum mechanics: the superposition principle. At the root of all quantum phenomena lies the notion of spacetime paths. It is with paths that both the fabric of four-space can be woven and the superposition principle of quantum mechanics be expressed. Most straightforwardly the observer can start by choosing a few (material) objects and by measuring the paths these objects take to map out properties of space through time. While doing so for smaller and smaller objects and decreasing scales, he/she encounters the distinct paths that quantum particles travel along. For example, when he/she performs the well known double slit experiment and identifies interference effects because electrons follow a multitude of routes. At this stage, the observer can start to quantify the properties of his/her spacetime by identifying different classes of paths. e.g., one class of paths contains all those routes that go through slit one and another class those that go through slit two. Of course, the two classes of paths are a priori a consequence of matter (the double slit), but the observer decides to extend his experiments all the way to the Planck scale to explore the very spirit of superposition.

On the Planck scale, geometric fluctuations in spacetime are large and inhibit him/her to perform accurate measurements. This does not surprise the observer because of the fundamental measurement uncertainties $l_P$ and $t_P$, as well as the vast energy $m_Pc^2$ that the observer needs by now to perform his/her experiments. However, these fluctuations do not preserve the shape of spacetime, while the observer wants to measure what paths comprise spacetime despite any such quantum uncertainty in geometry.

The observer then recalls the double slit experiment, as well as the entire first part of this monograph, with the two distinct classes of paths. He/she realizes that one can still measure paths that cannot be obtained from each other by simply changing the shape of space. i.e., paths are detectable despite measurement uncertainty if they cannot be transformed into each other through a homeomorphism of spacetime. Indeed, these are the topological obstructions ($S^1$ loops), in the form of three-tori and wormholes, that allow one to measure paths irrespective of quantum uncertainty in shape. Consequently, the properties of such (classes of) paths can be assessed by the observer, also when they are present on scales much larger than $l_P$.

However, when taking a good look at his/her, by now formidable, measuring apparatus, it is also obvious to the observer that he/she is as much observer as actor in these experiments. The quantum uncertainty in geometry that plagues him/her, is caused by his/her apparatus as a consequence of the observer-observed entanglement discussed earlier. i.e., to observe is to alter.

While performing experiments in this manner, the observer then develops the following insight.

It is through the observational comparison of paths that one comes to an assessment of their identity, associated with either an observer or observed, and thus the structure of dynamical quantum spacetime. After all, four-space paths constitute histories of individual entities as well as their interactions with others. Since all parties involved can equivalently act as both observer and observed, their histories and choices only have meaning with respect to each other. They must therefore also collectively *induce* each other to have any individual existence in the first place. Indeed, no path as a chain of events can remain to exist on its own when it is scrutinized and thereby changed. This while the joint existence of such observed-observed histories requires them to always be equivalent locally.

In other words, observers and observed are both distinct and equivalent throughout four-space. Hence, to carry the quantum entanglement of all paths connecting all observers and all observed in four-space, quantum histories can only enjoy local dynamical changes in the form of globally distinct *mimics* of any four-space path. i.e., it is the very act of path identification, so of local quantum observation, that requires paths to globally multiply in four-space by mimicry.

In this, an act of mimicry equates to an observer's four-space history choice that is globally distinct yet locally equivalent to an observed's history. Local refers to individual events here, while global pertains to all events in a history. The notion of dynamical mimicry therefore implies that four-space paths multiply through acts of observation, if an observer and observed are to express quantum spacetime globally in a tangible manner. In a poetic sense, observers and observed mimic each other in a dance of jointly carried existence. In plain terms, an object like a (small) electron or a (large) black hole can only enjoy meritful existence if there is always a distinct yet equivalent four-space entity to *compare to*. Indeed, in a quantum universe like ours one is never alone.

Pondering this some further, the observer realizes that he/she is only one entity in the universe and that Nature herself, as the collective of all particles and four-space regions, needs to take on the task of observationally identifying spacetime paths and the particles that travel along them. In doing so unremittingly, these very acts must induce persistent and dynamical changes in the structure of scrutinized quantum spacetime.

The observer also concludes that the nature of these multiplicative changes preserves the intrinsic topological properties of the original paths. e.g., consider a landscape with a bridge. One can follow many routes across it, so it is the connectivity of the bridge that matters, not its shape. Furthermore, when identifying that bridge, the interaction between observer and observed requires Nature to build at least one other bridge. If Nature would not, then no proper confirmation of the bridge's identity is possible at all. Because first cause lies in the act of identification, path multiplication is an expression of observation. Realize again in this that a path is a global entity, a history connecting four-space points A and B that can be arbitrarily far apart for an observer.

In all, quantum spacetime is multiply connected, built up by a plethora of topologically distinct paths that carry particles. These histories are caused by the acts of mimicry of all locally equivalent observers and observed, yielding a dynamical quantum spacetime that is locally but not globally causal. Such globally multiplied histories are a tangible expression of quantum spacetime, and their origin is hereby elevated to a fundamental principle: the mimicry principle (MP). At last, the very essence of quantum spacetime can be formulated.

## 4. The Mimicry Principle

The proposed paradigm is the following. Any reality requires individual observers and observed for its inception. This while everyone can be both an observer and an observed, enjoying each other's respective physical experiences in full. Consequently, the joint existence of observers and observed, as the collection of all physical phenomena, must derive from them mimicking each other unremittingly, if they are to be distinct yet equivalent. This yields a world of *identity by mimicry*. In it, any local stretch of a spacetime path/history results from the global induction of some new part of four-space, so in a topologically distinct manner. This mimicking idea is formulated in terms of a principle, through a single line of text:

**Any observer can mimic any observed's identity in terms of locally causal spacetime events, but only through the induction of four-space histories that are always globally distinct.**

The MP forms the heart of quantum spacetime and is the main message of this monograph. It has taken some pages to develop and the MP is used in the coming chapters to derive properties of, among others, black holes, dark energy, inflation and the Higgs boson. The new perspective is that any reality in quantum spacetime takes the form of observer-observed mimicry, with the collective of all such interactions multiplying the very entangled histories that weave the fabric of four-space.

According to the MP every perceived reality, so any observation, constitutes an act of mimicry because all observers and observed are locally equivalent. This while an observer and an observed should also be globally distinct, so when perceived as entire four-space histories of occurrences, for either to have an identity. All observers and observed are therefore both globally distinct and locally equivalent, and their joint mimicking drives existence in a self-inducing manner. The MP constitutes a quite interesting world, so it is prudent to contemplate it further. Before doing so, one can formulate a rhyme to sum up the essence of the MP in a non-technical sense. This to aid one's understanding.

*If there is you and there is me, then neither of us knows what either can be. But if you mimic me and I mimic you, then both of us know what always to do.*

So the moral, if not the exact physical content, of the story is the following. An entity's history cannot be one path or another, it must be one path and another. In this, mimicry yields a reality that is only locally and not globally equivalent. As such, readers need not worry in a philosophical sense about the absence of free will or creativity when the MP is applied to everyday life. On the contrary, the MP does not impede such notions. It actually facilitates them because mimicry introduces an unbound multitude of globally distinct four-space histories.

What one should conceive of as identity in the MP can be learned from the equivalence principle. The observer on Earth and the one being pulled by the rocket undergo the same physical experience. i.e., they feel the same forces acting. So it is the experiences of the observer and observed that matter in the MP, and they should be exchangeable. As stated before in the context of Schrödinger's cat, anyone should be able to take the cat's position and be subject to the same laws of Nature. Furthermore, if one thinks of intrinsic properties like spin $\frac{1}{2}$ and a black hole's event horizon, then it is the history of experiences *resulting* from these properties that defines an entity's identity. So it is the resulting history that can be mimicked. Inevitably then, the joint existence of observers and observed renders experience and observation globally distinct yet locally equivalent as well. Again, any local observation is indeed a global act of mimicry.

It is therefore the global aspect that is crucial in the MP and any individual local observation induces a distinct four-space history in quantum spacetime. Whether one studies an atom or detects the motion of a star or ponders the universe, all such actions constitute observations. These multiplied histories are in turn fully subjected to the MP. So the observational identification of all paths requires their very multiplication unremittingly, if they are to exist, and a Machian perspective on the universe results. In a physical sense, this means that a quantum universe, as expressed by a multiply connected dynamical spacetime and subject to its own fundamental measurement uncertainty, can only support the very paths that it needs to weave the local fabric of four-space through a global operational process of neverending mimicry.

Consequently, also all individual histories of observers and observed mimic globally distinct and locally equivalent paths through spacetime. To realize this, paths must at least be comparable through local observation. Under this observational act of identification that the MP requires, no globally distinct path can therefore remain to exist individually. If it would, then an observer could mimic all events in that history of an observed and thus lose his/her identity, violating the MP. Instead, when scrutinized by any observer, four-space paths multiply and thus continue to build quantum spacetime collectively and globally. Figure 1 provides a pictorial display of the MP.

Conversely, and according to the MP, paths that are distinct under continuous deformation then constitute an observable property of any spacetime subject to fundamental measurement uncertainty. Furthermore, these topologically distinct paths express entangled histories of observers and observed in an equivalent manner. If the above global quantum spacetime can be constructed, and the remainder of this monograph shows how, then it can be used as a blueprint by which to mimic four-space paths. These paths then allow one to derive and compute (novel) physical effects related to black holes, particles and spacetime itself, as observable entities. Indeed, any mimicking observer route through such a quantum spacetime constitutes a pertinent reality.

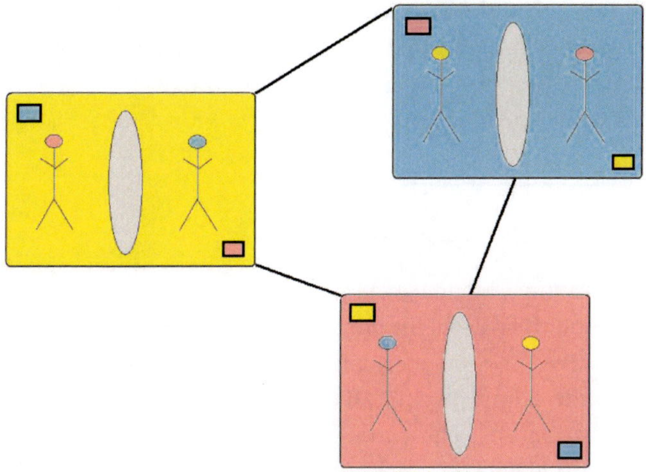

Figure 7.1. A pictorial description of the MP. The two stickmen are distinct (different colors) yet equivalent (same physiology) mimics. Each colored large rectangle represents a three-dimensional part of spacetime. These regions are connected by the black lines through three-dimensional space and by the different matching colored small rectangles (portals) through the fourth dimension of time, inducing entire histories (realities).

The above considerations are in line with the usual quantum phenomenology, in which no independent meaning can be assigned to observer and observed, and interactions between them affect both in a tangible manner. The four-space paths that connect observers and observed are now the fundamental objects in quantum spacetime. The MP allows Nature to distinguish such paths, despite quantum uncertainty in geometry, using topology. The MP also requires Nature to treat four-space paths as dynamical entities that express the simultaneous distinction yet equivalence of all observers and observed globally.

Finally, it is good to realize that the MP was already applied in a practical fashion to derive the temperature of a Schwarzschild black hole. Recall to this effect the rocket operator who mimics the effect of gravity under Newtonian conditions, so by being a distant observer along a locally equivalent path. This while the exchange of the necessary quantum entanglement information proceeds through a globally distinct $S^1$ loop, a closed path through time, that mimics how the observed-operator-acting-as-a-distant-observer performs a round-trip to the event horizon and back. With the MP in place, it is time to consider its consequences and applications. Who knows what more one can learn from it.

## 4.1. Everett's Many Worlds

The reader may wonder what the relation of the MP is to the many-worlds interpretation of Everett. In the latter, there is an observer-independent existence for the wave function. i.e., any given observer-observed subsystem that yields an outcome, simply splits off as an independent reality. This is called wave function collapse. The wave function associated with Everett's interpretation of quantum mechanics thus becomes multi-branched. In it, all branches or histories are non-interacting. According to the MP, observers and observed are equivalent, and there is no so-called wave function collapse to yield a specific observer defined reality. However, the MP also states that observers and observed only exist relative to each other. There is thus a new perspective that quantum spacetime as formulated in this monograph offers with respect to other studies.

Any particular choice for a history through four-space that a local observer-observed interaction may represent, is actually the expression of a world comprising dynamical four-space paths that derive their identity from each other in a global and multiplicative manner. The MP therefore renders all induced quantum histories fully *interactive*, both microscopically and macroscopically, rather than independent of each other. This is good to keep in one's mind. As mentioned before, time itself becomes multiply connected under the MP and no histories are ever truly disjoint.

## 4.2. A Few Mathematical Remarks

Being somewhat mathematically inclined, the observer in the MP thought experiment above can also recognize the importance of distinction when speaking of paths and their multiplicity in a formal sense. Path distinction is the expression of homeomorphic inequivalence. i.e., using the symbol $\sim$ for equivalence in this subsection, paths $X$ and $Y$ that obey $X \sim Y$ belong to the same class

$[X]$ because a continuous map transforms one into the other. Hence, to assert that there are paths $[X]$ the observer concludes that he/she must have at least one other class $[X']$ to compare to. Consider the landscape with a bridge again. One can walk along many paths $X$ in $[X]$ across the bridge and these paths are all topologically distinct from those paths $Z$ in $[Z]$ that choose to cross the water. The comparison class $[X'] \neq [X]$ that the MP causes through multiplication then simply represents all those paths $X'$ that cross *another* bridge.

In a mathematical sense this means that the minimum combination of paths is the loop. The attentive reader may recall in this the discussion on the squared modulus $\psi\psi^*$ as a probability density. For the wave function $\psi$ the origin of the square lies in the omnipresence of closed paths, so $S^1$ loops. When determining the black hole temperature, a loop (round-trip) topology is again induced to provide the rocket operator with the necessary information as he/she mimics the observed event horizon as an expression of the MP.

Furthermore, the topological dynamics approach of quantum spacetime can only involve integer numbers because distinct paths, carried by the portals through time (so the $S^1$ reality inductors that exchange observers and observed) of the lattice of three-tori with wormholes, are counted in such a way. This implies that whatever non-integer quantities one has, like mass or density, these must derive from $l_P$, $t_P$ and $m_P$ combined with scales that are determined in a global manner only.

## 4.3. The Lattice of Three-Tori Again

The motion of matter derives from a superposition of wave amplitudes which are determined by the actions associated with the followed paths. Recall in this the action path integral and how Feynman's perspective finds a natural implementation in the lattice of three-tori. The latter provides different topological classes of paths through the connected sum (also called three-ball surgery) introduced earlier. This procedure actually results in four topologically inequivalent paths associated with any three-torus, as a pertinent expression of multiplication (so induction) through the MP, as follows.

Consider a single three-torus $T^3$. As one travels for about a distance $l_P$ during a time $t_P$ from A to B, there are three internal $S^1$ loops to circle around, in compliance with Nature's measurement uncertainty expressed in terms of homeomorphically distinct temporal obstructions along three spatial directions.

One can also consider the simply connected (so non-loop) path that results from three-ball surgery. i.e., the beam that connects to an adjacent three-torus and does not cross the parts of the $T^3$ that are topologically identified through time. In all, one then finds four topologically distinct paths associated with any Planckian volume $l_P^3$ and any time interval $t_P$.

Of course, the MP turns this heuristic approach around. When one imagines a traveling particle and its observer, then the latter can choose to undergo the same experience every stretch of the particle's trajectory. Because the observer must maintain his/her own identity in this, an $S^1$ loop is induced that locally resembles the particle's path but not globally. Clearly, and as the simplest expression of the MP, a loop comprises two globally distinct yet locally equivalent paths between spacetime points A and B. Indeed, locally a circle looks like a line segment. Furthemore, any particle can move along the $x$, $y$ and $z$ directions. This in turn implies three $S^1$ loops for an observer under the MP, and a three-torus results. The same MP logic then applies to each three-torus as an observed entity connected to a mimicking observer. This must yield a lattice of multiplied three-tori in four-space.

Therefore, Planck scale quantum spacetime has $e_{\rm MP} = 4$ topological degrees of freedom corresponding to local patches of Euclidean space that are turned into the lattice of three-tori by the MP. Indeed, under the MP one goes from a local particle trajectory to a non-local loop and then to a global lattice of three-tori embedded in spacetime, with four topologically distinct paths connecting any two points A and B. Interestingly enough, Einstein gravity based on the local equivalence principle also has four degrees of freedom, as does Maxwell's electromagnetic theory of matter and light.

## 4.4. Black Hole Entropy

With the above MP perspective, one can explore black hole properties some more. A black hole is a thermodynamic object that emits like a black body at a temperature $T_{\rm BH} = \frac{g}{2\pi}$ for the surface gravity $g = 1/(4M)$, as found above. One also has that the black hole entropy is given by $S_{\rm BH} = CA$ with $A$ the surface area of the black hole, in the usual natural units. In this, $C$ is an unknown constant to be found. The surface area $A$ of a black hole is $A = 4\pi R_S^2$ for a Schwarzschild black hole with zero charge and zero angular momentum. The entropy expression holds because $S_{\rm BH}$ represents the total amount of in-

formation about the black hole interior that can be accessed from the outside. Indeed, the enclosing $S^2$ event horizon is all that an observer and observed has to work with, and therefore to mimic, when describing a quantum black hole. The notion that an enclosing surface constitutes a measure of total system information is usually referred to as the holography principle. i.e., the boundary of a system, the $S^2$ surface here, contains all the information necessary to define the system's internal quantum states.

More importantly, one finds that $C = 1/e_{\rm MP}$ when the MP is applied, as follows. Whatever information an observer can obtain from a black hole horizon, he/she does so by mimicking paths through quantum spacetime. Consider again the round-trip that leads to the black hole temperature $T_{\rm BH}$ of some infinitesimal surface area $l_P^2$ of the event horizon. The lattice of three-tori then also provides $e_{\rm MP} = 4$ paths *through* any such section of $S^2$ to connect it with the black holes's internal quantum states.

The crucial point here is that these four paths are globally distinct yet locally equivalent because the MP pertains to any evaporative signal through the leaky event horizon detected by a distant observer. Consequently, the information that is mimicked by a distant observer is quadrupally multiplied between the observed event horizon and the distant observer. The net entropy of a black hole's total surface area is therefore only $\frac{1}{4}$ of the total number $A/l_P^2$ of comprising Planckian areas. This is a fascinating result. Indeed, quantum spacetime mimicry leads to a four-fold *redundancy* in locally observable information.

Of course, $T_{\rm BH}$ and $S_{\rm BH}$ are thermodynamic quantities. They are related as such through the first law of thermodynamics, $T_{\rm BH}dS_{\rm BH} = dE - dW$. This holds for the black hole's increase in entropy $dS_{\rm BH}$, the increase in energy $dE$, and the exterior work done $dW$. In fact, it is easy to see that that the expressions for $T_{\rm BH}$ and $S_{\rm BH}$ are completely consistent with each other thermodynamically, for any black hole described by a mass, a charge and an angular momentum. As an example to show this, consider the case that $dW = 0$. Substitute $T_{\rm BH} = 1/(8\pi M)$ and $S_{\rm BH} = A/4 = (4\pi R_S^2)/4$, for the Schwarzschild case of $R_S = 2M$, into the first law of thermodynamics. In this, realize that a change in a black hole's energy equates to $dE = dM$. This while $dS_{\rm BH} = 8\pi M dM$. Hence, $T_{\rm BH}dS_{\rm BH} = dM = dE$, in proper order.

However, the point of this subsection is to highlight that the expressions for $T_{\rm BH}$ and $S_{\rm BH}$ can be derived independent of each other through the lattice of three-tori and the MP. Indeed, the $S^1$ loop used by the observed-operator-acting-as-a-distant-observer to mimic a round-trip to the event horizon and back also automatically yields the normalization $2\pi$ in $T_{\rm BH}$. This while the normalization of 4 in $S_{\rm BH}$ follows directly from application of the MP to a particle traveling through four-space. In other words, $T_{\rm BH}$ and $S_{\rm BH}$ are above all quantum expressions of spacetime, which together allow the MP to actually impose a first law of thermodynamics for black holes.

Once more one is confronted with the fact that leaky black holes can exist on all scales, coupling to small or large $S^1$ loops that mimic local stretches of spacetime. Quantum black holes thereby form an inescapable link between microscopic and macroscopic physics, through the lattice of three-tori. The remainder of this monograph pursues such a notion in depth and shows that not only the lattice of three-tori but also Wheeler's quantum foam is an expression of the MP.

# Chapter 8

# Topological Quantization of Spacetime

With the MP finally available to understand the nature of quantum spacetime, it is time to actually construct quantum four-space self-consistently through the already encountered building blocks, three-tori and wormholes. Because the latter elements are discrete, one can speak immediately of a *quantization* of spacetime. This chapter then finds its continued inspiration in Wheeler's quantum foam and Feynman's path integral formulation for quantum superposition, and uses topology as the natural language of quantum spacetime. The sections below present an intuitive construction of quantum spacetime based on the MP and its global observer-observed perspective. Subsequently, the equation of motion for the topological dynamics of quantum spacetime is derived, and solved under certain approximations to explore the nature of dark energy and its relation to inflation.

## 1. Topological Dynamics: Multiplying Wormholes

The MP implies that a single path modulo homeomorphisms (so continuous deformations) has the same physical significance as no path. Or equivalently, that the minimal building block of spacetime is the loop $S^1$. As should be clear by now, a loop is the combination of a pair of mimicking paths from A to B such that one cannot be deformed into the other by homeomorphisms: no entity has a meaningful existence only of itself, mimicking is mandatory for there to be any reality.

To be precise regarding the construction of quantum spacetime, some more mathematics are needed at this time. When dealing with four-space, the natural building blocks are so-called *prime* three-manifolds and their number evolution in time. This is because any topologically non-trivial spacetime manifold can be constructed by adding together three-primes, embedded in four dimensions, through the earlier encountered connected sum (so by using three-ball surgery). These three-primes, e.g., the handle $S^1 \times S^2$ and three-torus $T^3$ introduced above, lend their name from the property that they themselves cannot be written as the connected sum of other three-dimensional entities. Indeed, three-primes are the topological analogue of prime numbers in that they cannot be divided up further in three dimensions.

Furthermore, interactions among prime manifolds can never change their intrinsic properties, like the number of closed loops that they contain. After all, thinking back about the central lattice of three-tori with wormholes, one may imagine that three-tori and handles are added or removed. Still, there is no such thing as half a three-torus or half a leaky black hole. This being said, three-primes are more than a mathematical convenience. Indeed, even though loops constitute the basic elements in quantum spacetime, the evolution of our universe is undeniably one of three-dimensional space changing with time. Hence, to construct a proper four-space, $S^1$ loops have to be the temporally connecting elements that turn a sequence of three-dimensional spaces into an actual four-dimensional quantum history.

Indeed, and returning to physics, the MP (so multiplication) applies to any globally distinct yet locally equivalent path that an observer uses for his/her mimicry. Such a path is a four-space history and can pertain to a point particle. However, spacetime itself comprises Planckian volumes $l_P^3$ connected every time step $t_P$. These four-space regions are themselves also subject to the MP, as observers occupy them. In fact, any Planckian volume must contain *infinitely* many one-dimensional paths to build up a continuum. Hence, a history describing quantum spacetime is a four-dimensional embedding of a sequence of three-dimensional objects, each built from a continuum of loop-based (so topologically distinct) paths that constitute temporal obstructions. Quite a mouth full and a challenge to one's imagination.

Nevertheless, there is clear structure: three-tori and wormholes result when one applies the MP to, respectively, particles moving through four-space and black holes. Indeed, the lattice of three-tori, as the carrier of Feynman's path integral, was shown above to follow from the MP. Through the lattice of three-tori non-quantum macroscopic black holes turn into wormholes. However, it would be very satisfying if Wheeler's quantum foam of mini wormholes also follows from application of the MP.

To this effect, first recall that mini wormholes and macroscopic black holes enjoy Hawking evaporation and merging. Furthermore, mini wormholes and leaky black holes are both the carriers of Mach's global principle of inertia, on microscopic and macroscopic scales respectively. As such, their only difference lies in mass, i.e., geometry, and not in connectivity, i.e., topology. When one applies the MP with this perspective, then it follows that any leaky black hole can be mimicked by an object that is globally distinct yet locally equivalent. Such an entity can only be another wormhole, if it is to constitute identity by mimicry. In this, gravitational quantum uncertainty on the Planck scale is characterized by mini wormholes of mass $m_P$ and lifetime $t_P$. Combined, the MP then leads to the inevitable conclusion that Wheeler's quantum foam is a multiplicative phenomenon: a collective of transient Planck mass wormholes that are unremittingly induced by macroscopic black holes through the act of observation, so by identification!

It is important to let all the above sink in, for the right mindset to construct a dynamical quantum spacetime. The latter comes about because a collective of three-tori and wormholes allows *entire* quantum four-space histories to be constructed as globally distinct yet locally equivalent paths that are built from Planckian volumes topologically identified through time. Given that each multiplying three-torus can also store information on fermionic and bosonic particles, the MP allows for the creation of complete and interacting material realities through mimicry. This with transient mini wormholes induced by macroscopic black holes to carry the vacuum of quantum spacetime as perceived by Wheeler and the notion of inertia as perceived by Mach. This ever-changing world is the most that is asked from the reader's imaginative powers in four dimensions.

# 2. Construction of Quantum Spacetime through the Connected Sum

To realize the above, the MP dictates that the topological dynamics of any four-dimensional quantum spacetime must be carried by multiplication of three-primes and wormholes that contain portals through time in the form of $S^1$ loops, so pairs of mimicking observers and observed that induce reality. If one repeats this message of multiplication time after time, then the quantum spacetime constructed below follows, with the equations describing its dynamics. To be somewhat more rigorous mathematically, non-experts will have to deal with several technical remarks immediately below.

For example, the Euler characteristic of $T^3$ is zero, so it is the boundary of a Lorentzian four-manifold, which is basically a fancy name for a local Minkowski spacetime. This follows because the Euler characteristic of an $S^1$ loop is zero and the Euler characteristic of the three-torus is the three-fold product of $S^1$, so also zero.[1] Furthermore, the Einstein equation of general relativity enjoys the Schwarzschild solution for a black hole. The same holds for more general black hole solutions like the Kerr metric for a rotating black hole. Both these correspond to a handle topology, as argued in chapter 5. The Euler characteristic of $S^1 \times S^2$ is zero, so that it too is the boundary of a Lorentzian four-manifold.

Expert readers will know the mathematical form of the Einstein equation, with spacetime curvature proportional to matter density in the form of the energy-momentum tensor, but for this monograph it is only necessary to appreciate that all mass and energy follow curved four-space and may collapse to form black holes. Furthermore, one can consider the wormhole as a bridge between entirely different universes, with orthogonal sets of dimensions. However, a single multiply connected four-space is considered to be the universe here, albeit with many superposed quantum histories as described above (and further below).

A final technical remark is that there is one more three-prime needed. When one counts pairs of distinct paths, there should be the possibility of finding zero.

[1]For polyhedra the Euler characteristic $\chi$ is equal to the number of vertices v minus the number of edges e plus the number of faces f, i.e., $\chi$=v-e+f. Also, an $S^1$ loop is basically a triangle, so e=v=3 and f=0.

One may argue that this is not a physical situation. However, every loop locally resembles a line segment, while three-tori and black holes are embedded in a four-space that is locally Minkowski. With the loop being the central object of a multiply connected spacetime, the number *zero* represents the absence of temporal obstructions/portals. That is, one has a locally simply connected three-dimensional space for about a Planck time and the MP induces globally distinct yet locally equivalent paths starting from this null baseline. Such a zero loop state calls for the three-sphere $S^3$, embedded in four dimensions. The three-sphere is the counterpart of $S^2$ in one dimension higher. Hence, $S^3$ also has no boundary. It is easy to see that any closed loop in $S^3$, like circles on the surface of a sphere, can be contracted to a point because there are no obstructions. Indeed, the three-sphere is the zero element in topological dynamics.

With the basis set of three-primes in place, it is necessary to apply the MP. Starting with $T^3$, a spacetime with a single three-torus cannot remain to exist by itself. Rather, distinct yet equivalent identities must follow from the act of mimicry, as observers mimic observed particles happily traveling through local $S^3$ regions and thus induce temporal $S^1$ loops for each of the three spatial directions. This amounts to the creation of a novel three-torus, as explained before, and the discussion below is a slight rewording.

Even though embedded in four-space, $T^3$ is a three-prime and its paths in three-dimensional space proceed through the six faces that it possesses. Hence, the multiplication of three-tori that the MP requires, can only connect the additional three-tori through beams homeomorphic $S^3$. In other words, one ends up with the lattice of three-tori as derived on heuristic grounds above. The lattice of three-tori is thus constructed by performing a connected sum between any three-torus and its mimic, for all mimics. i.e., using three-ball surgery each three-torus is connected to another three-torus through a beam. Mathematically, a lattice $L$ of three-tori can be written as

$$L(T^3) = \oplus^j T^3_j, \tag{8.1}$$

where $\oplus$ denotes a connected sum. The Einstein summation convention (sum over repeated indices) is adopted again and the integer $j$ can be arbitrarily large unless noted otherwise. It is important to realize that the origin of $\oplus^j T^3_j$ in equation (8.1) lies purely in the dynamical consequences that application of the MP has for topology. That is, only connected pairs of three-tori have comparative meaning, as an entangled four-space observer and observed.

The construction of the lattice can be performed along each of the six directions defined by the six faces of a three-torus. Subsequently, one can always apply a homeomorphism $\mathbf{H}$ that relocates the beams between three-tori. These continuous deformations express quantum measurement uncertainty. Hence, for $j >> 1$ and all homeomorphisms $\mathbf{H}_k\{L(T^3)\}$ with $k = 1, \infty$, the equivalence classes $[\mathbf{H}_k\{L(T^3)\}]_p$ contain all topologically distinct paths $[p]$ on $L(T^3)$ under homeomorphisms $\mathbf{H}_k$.

This sounds very mathematical, but the above paragraph simply states that the paths created under the MP are topologically distinct. So they persist as one changes only the shape of spacetime, nothing more than that. Einstein gravity that bends $L(T^3)$ on large scales is thus naturally allowed by a subset of smooth homeomorphisms, as already argued before, with a particular curvature given by local mass density. As one zooms in on the Planck scale, Wheeler's quantum foam of mini wormholes dominates, as is further discussed below.

Through $L(T^3)$, the motion of matter and its associated action yields the superposition of wave amplitudes along many different paths. One thus obtains, from the MP and expressed by the topology of spacetime, the superposition principle as formulated by the Feynman path integral. Many of the paths that enter the Feynman path integral are continuous, but not necessarily differentiable (so smooth). This is in the spirit of the homeomorphic invariance that underlies the approach in this work, path connectivity is all that matters. The constructed distinct four-space paths are now expressed, independent of quantum uncertainty in geometry, by the global multiply connected character of quantum spacetime. Furthermore, handles (so black holes/wormholes) can be attached to the lattice of three-tori through the connected sum as well. This structure is denoted by $L(T^3)^+$,

$$L(T^3)^+ = L(T^3) \oplus \; [\oplus^j (S^1 \times S^2)_j], \tag{8.2}$$

for the arbitrary index $j$ now counting the number of quantum black holes. Because of the homeomorphic freedom of the connected sum, handles can connect to three-tori as well as to other handles. i.e., one can just move them around. The three-sphere $S^3$ acts as simply connected *glue* and one has that

$$T^3 \oplus S^3 \sim T^3, \tag{8.3}$$

and

$$S^1 \times S^2 \oplus S^3 \sim S^1 \times S^2. \tag{8.4}$$

With this first application of the MP, it is already possible to construct a continuous quantum spacetime from three-primes that possesses dynamically active global paths based on connected $S^1$ loops. That is, the MP acts even as one reads this... From the here and now to all observers and observed through-out spacetime, the MP unremittingly multiplies existence and creates multitudes of novel four-space histories. This is known to the reader from the chapters preceding the formulation of the MP of course, but now one may dig deeper, as follows.

Information on the number of connected three-tori and handles, as well as the local ($N$ bits) and global ($N'$ bits) information on particles that the $S^1$ loops comprise, is distributed through homeomorphisms. That is, the number of three-tori and quantum black holes with their observer-observed entanglement are global, so Machian, quantities. Knowledge of these integer numbers is accessible to all observers since it is a direct consequence of mimicry. This while acquisition of such information proceeds without violation of causality because the topological identifications in the lattice of three-tori allow any two quantum spacetime points to be infinitesimally close. The latter point deserves further elaboration.

## 3. The Wave Function of the Universe

Indeed, one may wonder how an observer, as a local expression of Nature, can physically know about the formation of a quantum black hole far away from him/her or about the entanglement of two particle histories in the lattice of three-tori. In classical spacetime, one can only see past events and their future consequences combined into one well ordered timeline. In quantum spacetime this is augmented with the notion of superposition. Taking a global perspective on quantum events then leads to a wave function of the universe $\Psi_U$, and one that enjoys correlations spanning across space and time because of the MP.

Hence, the formation of a quantum black hole can be seen as a history within $\Psi_U$ that comprises one extra leaky black hole, when so triggered by an observer, compared to other histories. The universe as a whole, and other observers in it, then occupies that new state in the present time of the aforementioned observer. This while before, the universe was in a superposition of states that involved both the successful and unsuccessful creation of the quantum black hole. In this, and following the MP, one should not speak of wave function collapse here (like in the Everett interpretation) because in quantum spacetime all mimicked paths co-exist and interact.

i.e., all the chains of events, so all the spacetime paths that lead to a particular present time in some reference frame, can be distinguished topologically on $L(T^3)^+$, and $\Psi_U$ comprises them all. Specifically, $\Psi_U$ is globally accessible for all observers in terms of countable quantities like three-tori, wormholes and the four-space histories carried by them. After all, the nature of the information involved in the creation of some entanglement ($N'_O = 1 \rightarrow 2$) or the formation of a quantum black hole ($+1$ extra) is just a single bit. The global distribution of this information is a mere choice within $\Psi_U$, in favor of a particular four-space history by mimicking it through a local observation. Yes, the latter is as literal as one can perceive it: assessing the existence of a black hole through any interaction constitutes local mimicry and demands global distinction under the MP through a temporal $S^1$ loop. There is simply no escaping one's own influence on the cosmos.

There is a clear analogy here with the superposition of two electrons, in a joint up/down spin $\frac{1}{2}$ state, that are moved macroscopically far apart. Observation of one electron in an up state leads to the instantaneous implementation of the down state for the other electron, and vice versa. The reason is that no information needs to be exchanged causally upon observation. This has already been taken care of while creating the macroscopic superposition in the first place: through the MP two entire electron histories are created that match the local mimicking observer who is involved. i.e., who locally assesses the joint spin properties of two electrons and thereby mimics their information content. The resulting two paths are close together in quantum spacetime thanks to the topological identifications (portals) through time, so the induced temporal $S^1$ loops of $L(T^3)^+$ that render them distinct yet equivalent through the MP. The exchange of information between these two mimicking paths is therefore locally causal, as Einstein would have liked.

Similarly, the formation of a quantum black hole is a history, one of many, that ends with a quantum mechanically leaky event horizon. In three-dimensional space, at some moment in time, such an event causes local geometric disturbances that travel causally from its source. In four-space, the formation of a handle constitutes a topological change that globally distinguishes one time-like slice through four-space (so our universe) from another, by a single unit of information. It is any $S^1$ loop in $L(T^3)^+$ that allows for this distinction and the exchange of information.

Indeed, as mentioned before, experiments similar to the double slit set-up appear not to produce an interference pattern if a single bit of information labels a path, and Schrödinger's entanglement is destroyed. The MP explains this phenomenon globally as a choice among entire four-space histories that are the product of observer-observed mimicry. Simply put, in the double slit set-up the experimentalist tries to reproduce Nature. These very actions unavoidably induce distinct four-space histories of outcomes under the MP, one of which the observer will remember. These points are returned to in the remainder of this monograph, but the qualitative arguments are as stated here.

From a philosophical point of view, one can argue the following. Quantum spacetime comprises many entangled histories, microscopically *and* macroscopically, all the result of mimicry. Any observation constitutes an act of mimicry: as an observer perceives of the observable world, so the latter enjoys a perception of the former. Therefore, performing an observation by oneself equates to being globally connected to an altered universe, through a mimic of one's own choosing. In practice the universe appears not to change, but by the MP it must. This is a fascinating realization and worth exploring further (see the final chapter). That being said, it is good to remain grounded and first return to the construction of quantum spacetime.

# 4. An Equation of Motion for the Evolution of Quantum Spacetime

So far, the discussion has been mostly qualitative. To make quantitative progress, an equation of motion should be derived for the evolution of quantum spacetime and it must be solved to make *falsifiable* predictions. The key in the derivation of the equation of motion is 1) to incorporate the MP through proper counting of multiplied paths that are embodied by the basis set of three-primes; and 2) to embrace the discrete nature of a topologically based quantum spacetime all the way to time itself.

### 4.1. Right Hand Side

First, one recalls that the MP requires four-space paths that are globally distinct yet locally equivalent between any spacetime point A and spacetime point B. The minimal duo being the loop $S^1$. Second, one remembers that it is necessary to have the loop-containing prime three-manifolds $T^3$ and $S^1 \times S^2$. Third, one *counts* all loops in three-primes self-consistently using the MP, every Planck time.

An observer, or Nature in general, who wishes to identify a three-torus locally can do so while approaching a $T^3$ along any of six spatial directions (top, bottom; front, back; left, right). In order to identify the properties of the three-torus from some side, through observer-observed entanglement and within a Planck time, the MP requires an additional three-torus. The same holds for all neighboring directions, also within a Planck time. Combined, this implies that the complete identification of a three-torus under the MP requires six additional mimics, and thus a multiplication factor of seven.

For the wormholes, the same logic applies. Albeit that a quantum black hole constitutes a bridge (so a handle) and identification can proceed only on either side of the bridge. This yields two sides, spatially and temporally separated, from which to assess the nature of the handle. The MP then imposes two additional mimics, every Planck time, and thus a multiplication by a factor of three. One can formulate the numerical considerations above in terms of the actions of the differential counting operators $\delta_j$, which are the topological successors of the geometric location and momentum operators encountered in chapter 2. One has an index $j = 0$ for the three-torus, $j = 1$ for the handle, and $j = 2$ for the three-sphere, with $P$ denoting a three-prime. Consequently,

$$\delta_j P_j = \mu_j P_j, \tag{8.5}$$

for the three-prime manifolds $P_j$, with

$$\mu_0 = 6 \ \& \ \mu_1 = 2 \ \& \ \mu_2 = 0. \tag{8.6}$$

Therefore, a single wormhole (three-torus) becomes a triplet (septuplet) of wormholes (three-tori) under the MP. This re-confirms the emergence of a lattice of three-tori. However, black holes carry mass and are subject to Hawking evaporation. So, the MP imposes additional pairs of mini wormholes, i.e., Wheeler's

Planck scale quantum foam, with a success that depends on the longevity of existing leaky black holes.

## 4.2. Left Hand Side

For the left hand side of the equation of motion one first realizes that the MP deals with global properties, basically just countable numbers. Furthermore, as mentioned before, continuous deformations (so homeomorphisms) acting on $L(T^3)^+$ allow for causally close proximity between any two four-space histories locally. Hence, globally only a first order in time differential is needed for the left hand side, as with the Schrödinger equation. That is, for wave phenomena like light one expects a second order differential equation to assure the local causal propagation through spacetime of its solutions. This while changes in the connectivity of spacetime itself are intrinsically global, and first order in time pertains.

Furthermore, the nature of the time variable, which is addressed in more detail below, has to be taken as *discrete*. After all, because the mathematical basis of quantum spacetime derives from continuity, physically measurable quantities that are globally accessible to an observer are integer numbers only, in natural units. Indeed, when speaking of space, one does so in units of uncertainty as the Planck volume $l_P^3$ and the Planck length $l_P$, respectively $T^3$ and $S^1$. While for time the proper unit is the Planck time $t_P \approx 5.4 \times 10^{-44}$ s. A small number to everyday experience, but larger than zero nonetheless.

Also, when counting as the MP requires, the massless three-torus has its four-space uncertainty specified by $l_P$ and $t_P$. However, the wormhole carries mass. Therefore, the natural unit of mass for mini black holes that result from induction (multiplication) in the equation of motion, is the Planck mass $m_P \approx 2.2 \times 10^{-5}$ g. In all, despite being measures of uncertainty, the numbers $l_P$, $t_P$ and $m_P$ constitute the proper units to quantify Planckian spacetime. Our current universe is well defined statistically because it comprises so many of them. The counted numbers of three-tori and handles are denoted by the multiplicities $n_j$, and one has that

$$\Delta n_j P_j = n_j(m+1)P_j - n_j(m)P_j, \tag{8.7}$$

for the temporal changes in $n_j$, discrete times $(m+1)t_P$ and $mt_P$ in physical units, and integer steps $m = 0, 1, 2, \ldots$ .

### 4.3. Final Form

The correct connection between the left hand side and right hand side again involves the MP. Evolution in the number of three-primes constructing quantum spacetime is simply $\Delta n_j$. However, for the counting of three-primes on the right hand side only a single three-torus and wormhole have been considered, with $\mu_0 = 6$, $\mu_1 = 2$ and $\mu_2 = 0$. Of course, the act of mimicry is performed unremittingly. Therefore, the total multiplication that one incurs under the MP is $\delta_j n_j P_j$. This yields

$$\Delta n_j P_j = \delta_j n_j P_j = \mu_j n_j P_j. \tag{8.8}$$

Because the index $j$ already labels the three-primes $P_j$ on which the operators act, one can also suppress the appearance of $P_j$ for ease of notation and simply write

$$\Delta n_j = \delta_j n_j. \tag{8.9}$$

Note that there is no topologically dynamical role for the basis three-prime $S^3$, as would be expected since

$$\Delta n_2 = \delta_2 n_2 = \mu_2 n_2 = 0. \tag{8.10}$$

However, for the quantum black holes there are matter degrees of freedom that play a role as well. After all, the number of leaky black holes changes as they form by gravitational collapse, evaporate by quantum processes or merge when brought into close proximity. These processes are denoted by $\Delta F$ (formation), $\Delta E$ (evaporation) and $\Delta M$ (merging), respectively. Such effects change the multiplicity $n_1$, subject to the discrete time variable $m$. Going from one Planck time to the next, and adding the matter degrees of freedom terms to the right hand side, one thus finds the quantum spacetime equation of motion as

$$\Delta n_j = \delta_j n_j + \delta_{1j}(\Delta F - \Delta E - \Delta M), \tag{8.11}$$

with $\delta_{1j}$ the usual Kronecker delta, which is a discrete function that is zero unless $l = j$. Equivalently, and explicitly including the discrete time dependence, one has

$$n_j(m+1) = (\mu_j+1)n_j(m)+\delta_{1j}(\Delta F(m+)-\Delta E(m+)-\Delta M(m+)), \tag{8.12}$$

where $(m+)$ denotes the continuous time interval $(m, m+1)$. Proper accounting of $n_1(m)$ under formation, evaporation and merging is implied in the above expression. i.e., $\Delta F = 1$ means the formation of one black hole (hence the plus sign). Also, $\Delta E = 1$ means that one black hole has evaporated during one time step (hence the minus sign). While $\Delta M = 1$ means that two black holes have merged into one (minus sign again).[2]

It should be noted, and this is discussed further below, that $n_1(m)$ pertains to black holes of all masses, from the Planck mass to solar masses and more. This is particularly relevant for the $\Delta E(m+)$ term, which has an implicit and very steep $t_e \propto M^3$ dependence on black hole mass $M$. So one can best write $n_1(m)[\{M_k\}]$ for the masses $M_k$ that black holes $k = 1...n_1(m)$ have. For ease of writing again, this is typically suppressed. In any case, solutions to equation (8.12) describe the evolution of quantum spacetime in terms of the collective behavior of three-tori and wormholes. This is one of the goals that was aimed for. So it is prudent to explore the nature of equation (8.12) further and discover what interesting solutions it may have.

## 4.4. General Relativity and the Equation of Motion

Equation (8.12) appears quite simple, but this is slightly misleading because the matter degrees of freedom term after the Kronecker delta (so for $\delta_{1j} = 1$) affects the number of quantum black holes in a manner that potentially involves the full impact of general relativity. Indeed, the evolution of the wormhole multiplicity $n_1$ is in part determined by the formation, evaporation and merging of black holes in our universe. These processes depend on the cosmic evolution of mass density and spacetime curvature as described by Einstein gravity. After all, the occurrence of structure in the universe is intimately tied to the formation of massive stars, their black holes as stellar remnants, the subsequent dynamical interactions between those black holes, and the latter's accretion histories.

Therefore, the longevity and transience of all black holes comes into play through the right hand side of equation (8.12). As such, general relativity remains valid down to the Planck scale as long as care is taken to *extend* it with the effects of Wheeler's induced quantum foam and black hole evaporation[3] as

[2]Naked singularities, so black holes without an event horizon, may in principle form and these could yield $\Delta E = 0$ after evaporation.

[3]The computation of the resulting spectrum for black hole evaporation knows some subtleties,

well as information effects through $G' \neq G$. When so incorporated, the terms $\Delta F$, $\Delta E$ and $\Delta M$ can be computed as one would otherwise do when following the joint evolution of mass density and curvature in Einstein gravity with $G \rightarrow G'$. In this, one should allow particles to follow the full multiply connected structure of $L(T^3)^+(m)$ and thereby properly account for all possible routes in Feynman's path integral.

Quantum gravity, as a topological extension of general relativity in the form of equation (8.12), is now an expression of the global multiply connected nature of quantum spacetime. To this effect, the MP provides the link between the macroscopic and microscopic wormhole cosmos. Long-lived black holes (so with large values of $t_e$ cosmologically) induce mini wormholes on the Planck scale, as *miniature* copies of the large scale originals, and provide spacetime quantization in terms of mass. Indeed, the MP is always hard at work.

The *natural* initial condition for equation (8.12) is $\mathbf{Q} = 2T^3 \oplus 3S^1 \times S^2$. This is because the combination $\mathbf{Q}$ allows for a multiply connected four-space at $m \rightarrow 0$ that has as many topologically distinct paths connecting three-tori as it has connecting wormholes, namely $e_{\rm MP} = 4$. In this, first consider the beam connecting the two three-tori. Then also pass through the three wormholes or the three loops of either three-torus. Wormholes introduce a notion of mass and one would expect that the four degrees of freedom of Einstein gravity then derive self-consistently from both the lattice of three-tori (massless) and the attached wormholes (massive). So three-tori and wormholes should generate the same number of globally distinct yet locally equivalent paths between any points A and B. Furthermore, because quantum black holes evaporate, this naturally breaks the symmetry of $\mathbf{Q}$ and the evolution of our cosmos proceeds through extended general relativity as described above. When the universe is subsequently constructed through the MP, with $\mathbf{Q} \rightarrow L(T^3)^+$, there are no so-called island universes but only a fully path connected quantum four-space.

## 5. A Topological Extension of Mach's Principle

Before solving equation (8.12) under certain circumstances, there is an immediate result that can be read off in the context of Mach's principle, and that was already emphasized above. The quantum foam stability problem that plagues

but for equation (8.12) only the overall decrease in mass is relevant.

general relativity on the Planck scale, as explored by Wheeler, finds a natural resolution in the quantum spacetime equation of motion. If one considers the limit where

$$\Delta n_1 = \Delta F = \Delta M \approx 0, \tag{8.13}$$

so a static or very slowly evolving population of quantum black holes with little formation and merging due to matter and curvature degrees of freedom, then

$$2n_1(m) \approx \Delta E. \tag{8.14}$$

The quantum foam instability problem is therefore solved *if* $n_1(m)$ in equation (8.14) pertains to long-lived quantum black holes, so that $n_1(m)$ is approximately constant. i.e., if there is an ensemble of leaky black holes that live for discrete times $mt_P$ at least comparable to the current age of the universe, despite evaporation. This population $n_1(m)$ must then induce the evaporation $\Delta E$ of twice that many wormholes because quantum black holes are unremittingly mimicked (as Planckian miniatures) under the MP.

Mathematically, $n_1(m+1) \approx n_1(m)$ for $\Delta n_1 \approx 0$. So the evaporation term $\Delta E$ must act on not more than the Planck time for consistency. i.e., the induced evaporative term is truly transient. Physically, the MP shows its true force in that the existence of a leaky black hole, as an observable entity, requires it to be mimicked in two-fold through globally distinct yet locally equivalent wormhole pairs on the Planck scale. This to define the black hole's quantum identity every discrete time $mt_P$.

The natural mass for these induced wormholes is therefore the Planck mass, if they are to evaporate in a Planck time as argued before. This while the evaporation time of a black hole scales as its mass to the third power, $t_e \propto M^3$. So a black hole of more than $\sim 5 \times 10^{14}$ g would live as long as 10 billion years. Hence, an ensemble of macroscopic quantum black holes provides a naturally stable quantum foam of Planckian mini wormholes because of the MP and expressed as a solution to the equation of motion (8.12). This leads to the quantum foam that Wheeler envisioned, but one that is constrained by the longevity of macroscopic quantum black holes. In all, one has a very nice expression of how the MP connects the macroscopic and microscopic cosmos in terms of vacuum energy and inertia.

Indeed, Einstein's interpretation of Mach's principle states that the global distribution of matter somehow determines the (changes in) local geometry, and vice versa. Wheeler's local quantum foam of mini wormholes, with its multiply connected topology, is suggested by general relativity and its lack of invariance under conformal transformations. i.e., recall that smaller volumes lead to larger densities and thus curvature. Therefore, following Einstein's ideas for four-space geometry, with the MP taking care of quantum spacetime topology, matter that is globally observed to be in the form of macroscopic quantum black holes must somehow determine the (changes in) Planck scale topology, and vice versa. One may even go one step further and use the MP to derive the topological extension of Mach's principle (EMP). It is formulated as follows.

*The global topology of and distribution of matter in the universe determine the changes in Planckian topology and geometry, and vice versa.*

This EMP is in essence the answer to the following question. How does Nature know what quantum fluctuations to provide for, in every Planckian volume? The answer found here, is that the collective action of all long-lived quantum black holes provides this. Although strange, the EMP is in the spirit of Mach's and Einstein's ideas, with some help from the MP. Indeed, from the EMP (and the MP) the important conclusion from above on Planckian spacetime immediately follows: Wheeler's quantum foam is a collective of transient multiplying wormholes induced by macroscopic black holes through the act of observation.

The number[4] of macroscopic quantum black holes in the *observable* universe, so in all of three-dimensional space for some time slice through causally connected four-space, is therefore the global quantity that determines the local occurrence of mini wormholes in Wheeler's quantum foam. Because of homeomorphic invariance (the freedom to perform continuous deformations) one may think that there is no specific volume to which induced mini wormholes couple. i.e., one needs a volume $V$ to turn a pure number like $2n_1(m)$ into a mass density $2m_Pn_1(m)/V$ to let the Planckian vacuum gravitate. However, despite being global principles, the MP and the EMP together do constrain this volume, as follows.

To speak of a Machian universe in the extended topological sense, there should be a stable global topology and one that carries mass across time so that material objects have a well defined identity. In other words, one needs black holes to introduce mass globally. The inducing population of macroscopic

[4]Not the number density.

quantum black holes should therefore have a member whose longevity $m't_P$ exceeds the age $mt_P$ of the universe at that time. This $m' > m$ black hole lifetime is significant because the MP allows such a history to be mimicked, and thus a global notion of matter to be established into the future. i.e., with time being relative, any observer perspective that enjoys $m' > m$ for some long-lived quantum black hole, called black hole $X$ from now on, requires the entire causally connected universe to be mimicked...

Clearly, this introduces a special duality into quantum spacetime: mimicry also pertains to the universe itself, to establish its identity through the global entanglement of material observers and observed. Subsequent occurrences of such long-lived black holes do not carry the same significance because the process of creating two distinct yet equivalent versions of the evolving universe is simply repeated. If one can count from one to two, then subsequent numbers are easy. Furthermore, the MP dictates that equation (8.12) holds for all quantum black holes. This while any local observer of the quantum foam can determine the number of such macroscopic quantum black holes in the wave function of the universe $\Psi_U$. This renders the first occurrence of a black hole that exists for longer than the contemporary age of the universe a locally and globally observable quantity on $L(T^3)^+(m)$ in four-space, as follows.

Any observer must be able to mimic the vacuum energy density of Wheeler's induced quantum foam locally. Conversely, the volume $V$ that defines this mimicked vacuum mass density has to be properly defined globally, like $n_1(m)$ and $m_P$. This is only possible through the geometric size of the causally connected universe itself. So, if $L_f$ is the size of the causally connected cosmos when black hole $X$ forms, then $V = L_f^3$. Indeed, $L_f$ is the size of the universe that needs to be mimicked when black hole $X$ forms and thereby introduces a future with longevity for our cosmos in a distinct yet equivalent path sense. Note here the subtle role played by causality through observers and observed that enjoy a common local origin, but in the presence of the global quantum entanglement carried by the wave function of the universe.

The globally frozen in scale $L_f$, particularly its numerical value, is readdressed in the context of dark energy in the next chapter. For that reason, the vacuum energy density that Wheeler's induced quantum foam represents, is written as $\Lambda(m) = 2n_1(m)m_P/L_f^3$. This uses the common cosmological constant symbol $\Lambda$. The important point to realize is that despite the freedom of continuous deformations, the topological dynamics approach of paths can identify scales beyond $l_P$, $t_P$ and $m_P$.

## 6. Some Further Solutions: Exponential Self-Multiplication and Decay

Besides the Machian solution above, equation (8.12) can also be solved straightforwardly under other approximations. For example, if $\Delta F = \Delta E = \Delta M \approx 0$ then

$$n_j(m+1) - n_j(m) \approx \mu_j n_j(m), \tag{8.15}$$

with the approximate exponential solutions

$$n_0(m) = 2 \times 7^m \tag{8.16}$$

and

$$n_1(m) = 3 \times 3^m, \tag{8.17}$$

for initial conditions of the universe at $m = 0$ that involve **Q**. Hence, quantum spacetime can experience a self-multiplication phase that very quickly builds up a lattice of three-tori with a quantum foam of mini wormholes attached to it. Such a state for the quantum foam is re-addressed in the context of inflation in the next chapter. For the three-tori, there are no further solutions unless one allows for a reversal of the arrow of time. Although without empirical support, such a notion is popular in many research efforts. In this monograph, it could lead to island universes as three-tori are removed from $L(T^3)^+$. Of course, such a universe would constitute the very opposite of the spirit of the MP, and also have little longevity.

## 7. The Global Nature of Time: Counting Three-tori

The equation of motion derived above is discrete in time, but with continuous time intervals, because the MP contains the concept of identification. The latter is a discrete process, also when it comes to the fourth dimension itself. One may feel uncomfortable with this approach to the nature of time. However, it is possible to use the $T^3$ solution in equation (8.16), and the structure of $L(T^3)$, to provide a solid foundation for time as a global topological notion as well as an approximate geometric variable.

Unlike black holes, three-tori always increase in number monotonically as $n_0/2 = 7^m$ for $m \geq 0$. Turning this around, it is the topological number $n_0$ that is fundamental and globally known according to the MP. Hence, a sense of topological time is naturally defined as the quantity $m = {}^7log(n_0/2)$, or $mt_P$ in physical units. i.e., the MP endows continuous four-space with a countable three-dimensional space structure that all local observers can perceive through their mimicry. This should be no surprise by now.

It is not the case that every observer goes about the task of counting three-tori to experience some sense of time. After all, time is merely the parameter to denote the local changes that the universe undergoes from one's own perspective, without any absolute standing. This while the number $n_0$ is currently $n_0(m_0) \sim 7^{610}$, for an age $m_0$ of the universe of $\sim 14$ billion years (or $\sim 10^{61}t_P$).

Rather than counting the number $n_0(m_0)$ explicitly, one concludes that as many as $n_0(m_0)$ degrees of freedom are required to exhaust the state space capacity of the mimicking lattice of three-tori with occupying particles, and to expose its current fine-grained structure. With $\sim 10^{80}$ particles in the universe, the latter is not likely to occur and even a multiverse of many different, parallel yet interacting, histories can be supported by $L(T^3)^+(m_0)$, for every observer. Most fundamentally, it is the *ever increasing* amount of information that describes these histories that underlies any notion of time. This point is returned to in the final chapter.

The global notion of time explored above, based on counting, may seem at odds with the usual local relativistic nature of time. However, one actually complements the other, as follows. The number of three-tori, like the number of birds in the sky, is something all observers can agree on. Information on the number of birds can only travel at the speed of light, so has a local effect. On the Planck scale, as mentioned above, the lattice of three-tori enjoys continuous transformations (so homeomorphisms) as an expression of quantum uncertainty. These allow any two quantum spacetime points to be brought into close proximity, facilitating information transfer on the number of three-tori without exceeding the speed of light. The global count of three-tori is therefore a measure that identifies many interconnected four-space histories, with each of these spacetime paths comprising many observers that enjoy their own sense of local time.

Also, the MP preserves the feature of every $T^3$ that it is the boundary of a Lorentz four-manifold. Recall in this that the so-called Euler characteristic of a three-torus is zero. The same holds for the wormhole/handle $S^1 \times S^2$. Consequently, $L(T^3)^+(m)$ yields a Lorentzian signature $- + ++$ for the four-space that it bounds, with the local causal properties envisioned by Minkowski and obeying Einstein's relativity principle. So, as one would expect, for $m >> 1$ topological time $mt_P$ can behave just like the everyday notion of geometric time $t$, i.e., for scales much larger than the Planck scale. Nevertheless, the discrete and global nature of time also implies that there are many slices of it in, say, a second (as many as $10^{43}$ from our perspective). This while the topological identifications that connect these slices in $L(T^3)^+$ need not be infinitesimally close for all local observers. Hence, for a time interval $\Delta m >> 1$ one must speak of *composites* of partial four-space histories in the form of many time slices of a local Planck time wide that are connected macroscopically in four-space through $S^1$ loops.

Indeed, topological time only cares for a quantum four-space that comprises many multiply connected histories, all sharing a discrete identifier $m$. The latter therefore signifies how many distinct yet equivalent histories, $n_0/2 = 7^{m_0}$ to be precise, are accessible to a local observer whose clock indicates a time $m_0 t_P$ with respect to the big bang. Consequently, if $2\Delta m \leq m$ does not hold, then the coarse-grained nature of topological time becomes apparent. i.e., one starts to see the composites (so the multiple partial histories) connected to one's personal spacetime path. Indeed, the mimic of some $\Delta m$ composite should still be able to fit into four-space at topological time $m$. If it does not, so if $2\Delta m > m$, then its realization can only be part of an observer's possible future. And yes, this effect depends on the local notion of time $t$ that connects an observer to the temporal riches of $L(T^3)^+(m)$.

In a very real sense then, the possible histories of our universe are in superposition with themselves, and their composing and mimicked parts are thus accessible to observation. Looking back into the history of the universe, which is possible thanks to cosmic expansion, one should be able to see such time discreteness in the otherwise isotropic cosmic microwave background. This on a scale $\geq L_0/2$, with $L_0 = ct_0$ for $t_0$ the current age of the universe and $L_0$

the largest scale that can be observed today. For $2\Delta m > m_0$ the members of any pair of three-tori in $L(T^3)^+$ are macroscopically separated and topologically identified on a scale $L_0/2$. Through this connection between two three-tori cosmic microwave background radiation is temporally displaced over $t_0/2$. Because $L_0$ is what an observer perceives of the universe at $t_0$, geometric time only holds for spatial scales $2L < L_0$. Note here that quantum spacetime is still continuous. It is our perception thereof that becomes crude.

This coarse-grained character of geometric time therefore leads to a global anomaly, a dipole asymmetry in the cosmic microwave background's tiny temperature fluctuations. This topological redistribution process has some efficacy, which can be estimated as follows. As a consequence of the MP, each three-torus has a multiplication factor $F = 1 + \mu_0 = 7$, i.e., two septuplets of three-tori perturb the cosmic microwave sky. Hence, some amount of radiation on a scale $L_0/2$ of one three-torus is temporally displaced across a scale $L_0/2$ to the other three-torus, and vice versa. One thus finds for the amplitude $B$ of the resulting fluctuation in CMB power (equivalent to a lack of correlation because of the displacement) that $B = \frac{1}{2F} \approx 0.07$. Recent observational data indicate a large scale dipole asymmetry with an amplitude of $B = 0.07 \pm 0.02$. This empirical value of $B$ is suggestively close to the theoretical prediction. Furthermore, the Planck data indicate spatial scales of $L_0 - L_0/60$ for this feature, consistent with $L_0/2$ and its $\sim (l/L_0)$ times weaker mimics on smaller scales $l$.

Electromagnetic signals across time intervals $\Delta m \leq m_0/2$ experience no temporal displacement effects. But the early universe, more than 7 billion years ago, should also exhibit temporal displacement features. The distribution of luminous matter (e.g., galaxies) is the best place to look for these fluctuations. In a series of time slices through four-space, they emerge as shifted copies of the $L_0/2$ spatial scale components of our cosmos. Of course, the small amplitude $B$ of about 7% makes such coarse grained time fluctuations on a scale $L_0/2$ difficult to discern in an ensemble of large scale tracers like galaxies. In any case, even when it comes to time itself, the MP is what induces any notion of temporal existence. Rather than tempus fugit, so time flies, one can better say that time multiplies in the form of entire histories.

# Chapter 9

# Further Implications and Predictions

In order to test the ideas of topological dynamics in a definitive manner, pertinent predictions should be made. Indeed, this is the natural way that any theory should develop, through observations and experiments that allow it to be falsified or confirmed. Think in this light of the prediction that Einstein made on the bending of light by the curved spacetime around our Sun. This chapter offers a few more tests that can confirm or refute topological dynamics. Specifically, focus lies on dark energy, inflation, the growth of quantum black holes and the Higgs boson.

## 1. Dark Energy

The expansion of the universe appears to be accelerating, as indicated by the brightness evolution of so-called type Ia supernovae over the past $\sim 10$ billion years. Type Ia supernovae are evolved stars that explode in a binary configuration, through mass transfer between the companions. They act as standard candles, so are sources of an a priori known luminosity. This accelerated expansion of the cosmos is a *negative gravity* phenomenon and constitutes one of the biggest mysteries in cosmology, driving many observational and theoretical efforts. Dark energy appears to be fundamentally connected to the dynamics of quantum spacetime, as is explored in this section.

## 1.1. The Globally Embedded Quantum Foam of Wheeler

The evolution of the wormhole multiplicity discussed above leads to the unambiguous MP (and EMP) prediction that macroscopic leaky black holes induce pairs of mini wormholes, as transient miniature copies. Furthermore, the induced mini wormholes of equation (8.14) require an increase in the volume of three-dimensional space for their embedding in four dimensions. That is, the MP transforms a local simply connected three-dimensional region into a pair of mini wormholes under the act of quantum black hole observation. The tunnels that these induced mini wormholes represent have to connect with future four-space. Hence, there needs to be extra volume in the future to receive them. Consequently, three-dimensional space appears to be expanding to a local observer.

As a lower-dimensional analogy, imagine an $S^2$ world in the form of a sphere with its past internal to the closed surface. As one forms a tunnel through the third dimension, two areas in the sheet bulge out into the future and connect away from the locally flat plane. In effect, the two areas are pulled into the larger future volume surrounding $S^2$. Doing this for all local areas leads to a globally expanding $S^2$ universe. Indeed, our universe is locally $S^3$ and is pulled into a four-space future of larger volume by the actions of Wheeler's induced quantum foam.

One therefore immediately finds the correct sign for the (in)famous cosmological "constant" $\Lambda$ that Einstein introduced, and regretted so much. About a century ago common wisdom dictated a static universe, which general relativity did not allow for. To mend this problem, empty space was augmented with a notion to inflate, but the origin and magnitude of this negative gravity were not specified by general relativity. The latter was much to the dislike of Einstein. With the discovery of cosmic expansion by Hubble, the cosmological constant was abandoned by Einstein. It is said that he referred to it as his biggest blunder, to his colleague Gamow and in German, although history is unclear on this point. However, with the MP at the foundation of a quantum foam of induced mini wormholes, there is the *natural* ability for spacetime to expand through the presence of long-lived leaky black holes. In other words, this form of negative gravity has a topological origin.

One can now understand the origin of dark energy as induced mini wormholes in the following manner, pursuing the MP (and EMP) somewhat further yet again. Matter is quantized and expresses itself through the creation and destruction of particles. Similarly, spacetime is quantized and expresses itself through the creation and destruction of paths. Quantum black holes act as bridges through time in a four-dimensional sense, so their mass is spread out across space and time. They therefore induce mini wormhole pairs by number, as non-local and transient expressions of inertia, because the collective of all quantum black holes constitutes a global and long-lived notion of mass that local observers can detect and thus mimic. That is, stable quantum black holes provide a four-dimensional baseline for inertia on the Planck scale, in the proper form of distinct yet equivalent paths. This way, no a priori mass scale needs to be imposed on the vacuum of our universe: quantum black holes allow empty spacetime to carry mass. Because the universe can best be seen as a long-lived quantum fluctuation itself, this automatically allows for massive particles to emerge from the vacuum.

One may also reason from the MP directly. Because of the lattice of three-tori, an expression of the MP, the identity of black holes as wormholes is undefined for short durations (so within $t_P$). The observable local presence of an event horizon, on either side of the throat, therefore requires the unremitting presence of Planck-scale mimics for comparison. In this, induced mini wormholes appear randomly throughout a time slice $m$, so at any spatial position, because only the integer properties of the four-space histories in $\Psi_U$ matter. Either way, it is the global distribution of matter in the form of macroscopic black holes that determines local topological fluctuations (i.e., the EMP).

Furthermore, the number of macroscopic leaky black holes is a function of time and may vary from one time slice of thickness $t_P$ to the next. That is, $\Lambda$ need not be a constant in time, even though it is uniform in three-dimensional space. To be more quantitative about this, recall that the EMP leads to a specific scale $L_f$ equal to the size of the universe when the first black hole (called $X$) was formed that lives longer than the age of the universe at that time. With $N_{\rm BH}(m) = n_1(m)$ the number of macroscopic leaky black holes as a function of topological time, one finds that Wheeler's induced quantum equates to a dark energy density of

$$\Lambda(m) = 2\frac{N_{\rm BH}(m)\, m_P}{L_f^3}. \tag{9.1}$$

Note here that the topological time $m$ manifestly expresses that no violation in causality occurs as information on the total number of macroscopic black holes is distributed every Planck time through the MP, so across entire time-like slices of four-space, as an expression of quantum superposition. Time is still relative of course and one may freely choose any coordinate system, a so-called foliation, for the chain of time slices up to some $m$. In cosmology it is common to use the redshift $z$ for this, where $1 + z$ measures the change in the linear size of the expanding universe and $t/t_0 \approx (1+z)^{-1.5}$ gives the fractional age of the universe, for $t_0 = 14.3$ billion years today. So $N_{\mathrm{BH}}(z)$ is used in the remainder of this subsection.

Equation (9.1) shows that, for $N_{\mathrm{BH}}(z)$ independent of redshift, any local observer concludes that the dark energy density $\Lambda$ in g cm$^{-3}$ stays constant in time as the universe expands. i.e., a true cosmological constant is allowed for this topological form of dark energy, but this depends on the black hole number evolution of the universe. As the reader may recall, black holes are formed as a by-product of stellar evolution, and are thus tied to cosmic history.

## 1.2. Dark Energy and the Black Hole Formation History of the Universe

As is evident from above, the dark energy density of our universe is linearly proportional to the total number of macroscopic leaky black holes in three-dimensional space at any redshift labelling a time slice through four-space. The bulk of the stars, and thus stellar origin black holes, in the universe appears to be present as early as $z = 1$ (nine billion years ago), when a $(1+z)^4$ decline ensues in the cosmic star formation rate after a peak at $z \sim 2$ (eleven billion years ago). Note that this is an empirical fact, established through astronomical observations. Recall also that roughly 1% of all formed stars are so-called high-mass ($> 8M_{\odot}$) stars.

Furthermore, black holes are the remnants of such high-mass stars. One therefore expects to find an effectively constant $\Lambda$ for $z \leq 1$ because the bulk of all stars, and thus black holes, has been formed by then. Conversely, given the rapidity with which massive stars, leading to stellar origin black holes, are produced during $z = 1 - 2$, one expects a strong decrease in the number of macroscopic black holes, and thus in $\Lambda(z)$, for time earlier than $z = 2$. For a local observer today, it is plausible to consider $\Lambda(z)$ normalized to its current value $\Lambda(0)$, where the latter corresponds to the present number of black holes in the observable universe. The ratio $\Lambda(z)/\Lambda(0)$ can be expressed in terms of $dN_{\rm BH}/dz$, using astronomical observations and for a *co-moving* volume. The latter pertains to an observer moving with the overall cosmic expansion, and one has that

$$\Lambda(z)/\Lambda(0) = \frac{\int_y^z dN_{\rm BH}/dt\, dt}{\int_y^0 dN_{\rm BH}/dt\, dt} \approx (1+z)^{-0.4} \text{ for } z < 1. \tag{9.2}$$

This integral simply computes how many black holes have been formed in the past normalized to the present number. In this, macroscopic black hole formation starts at some large redshift $y > 30$ but at a very modest pace. Although a redshift coordinate is adopted for different time slices through four-space, in the form of $dt(z)$, there is no pertinent dependence on coordinate system because of the normalization in equation (9.2). Any observer, whatever its coordinate system, finds the dimensionless ratio of two numbers to be independent of the adopted coordinate system. It is assumed here that any observer can derive the relative temporal change in the total number of black holes, as three-dimensional space evolves, because he/she lives in a representative comoving volume that undergoes the same relative evolution in black hole formation as the observable universe.

One finds numerically that $\Lambda(z)/\Lambda(0)$ roughly scales as $(1+z)^{-0.4}$ for $z < 1$. i.e., $\Lambda(z)$ changes by about $\sim 30\%$ during the $z = 0-1$ epoch (about nine billion years into the past). The value of the so-called dark energy equation of state parameter, defined as $w = p/\rho$ for pressure $p$ and mass density $\rho$ is observationally constrained to $w \approx -1.1 \pm 6\%$ for $z < 1$, the latter through observations of type Ia supernovae as well as the cosmic microwave background, in particular by the Planck satellite. That satellite resided for several years in one of our solar system's Lagrange points (number two in fact) and made a high spatial resolution map of the cosmic millimeter sky. In any case, these observational data are consistent with equation (9.2). This is because a constant value of $w = -1.1$ is equivalent to a change in $\Lambda(z)/\Lambda(0)$ of about 30%, for $z < 1$. The latter follows from the conservation equation for an expanding universe (a solution of general relativity)

$$\frac{d\rho/dt}{\rho} = -3H_e(1+p/\rho) = -3H_e[1+w(z)], \tag{9.3}$$

with the Hubble parameter squared $H_e^2 = 8\pi G\rho/3$ basically the curvature. The latter represents the rate with which the universe expands (so $1/H_e$ is a time scale and called the Hubble time). If one expands $1 + w$ as $1 + w \approx \delta w$, around $w \approx -1$, then the relative change in $\rho$ over a local Hubble time is about $(H_e\rho)^{-1}d\rho/dt = -3\delta w$, for a constant variation $\delta w$.

Of course, there is observational uncertainty in deriving the black hole formation rate from the cosmic star formation history. In particular, one often adopts a Salpeter initial mass function. The latter refers to the name of the scientist who made a first model for the number of stars by mass, particularly those above 8 $M_\odot$ that yield stellar origin black holes. If this Salpeter initial mass function applies, then about 1% of all stars in the universe are more massive than 8 $M_\odot$, independent of time. Taking astrophysical observations into account, there can be a factor of a few uncertainty in this 1% number for early cosmic times, so for $z \geq 1 - 2$. Nevertheless, the above $w = -1.1$ behavior of the dark energy density persists for $z < 1$.

It is possible that many primordial black holes exist, comparable in number to the stellar origin black hole count. If the former happen to evaporate for $z < 3$, then values of $w$ larger than $-1$ are possible during that cosmic epoch since the dark energy density diminishes as $N_{\rm BH}(z)$ goes down with decreasing redshift $z$. This is because the MP does not care about the individual masses of quantum black holes that induce pairs of mini wormholes. Such lightweight primordial leaky black holes can be a result of the exponential self-multiplication discussed above. They only survive until today for masses larger than $\sim 5\times10^{14}$ g.

In all, the $(1 + z)^{-0.4}$ dependence of equation (9.2), so a $w = -1.1$ equation of state, is the most likely evolution of the dark energy density for modest redshifts. It is tantalizing that dark energy is directly coupled to the collective of macroscopic black holes. It is apparent that a small improvement in future measurement precision of an effective $w$ around $z \sim 1$ could confirm or rule out the topological dynamics model for dark energy advocated in this monograph. Furthermore, the one-to-one correspondence between the number of macroscopic black holes and the dark energy density is very much open to astrophysical detection, given the accuracy with which the cosmic star formation history can be probed. Indeed, for redshifts larger than $1 - 2$ stellar origin black holes quickly diminish in number and so should the induced mini wormhole dark energy.[1] If verified, then this one-to-one link constitutes the proverbial *smoking gun* in support of the validity of the MP (and EMP). Figure 2 presents the relative change in the dark energy density with redshift.

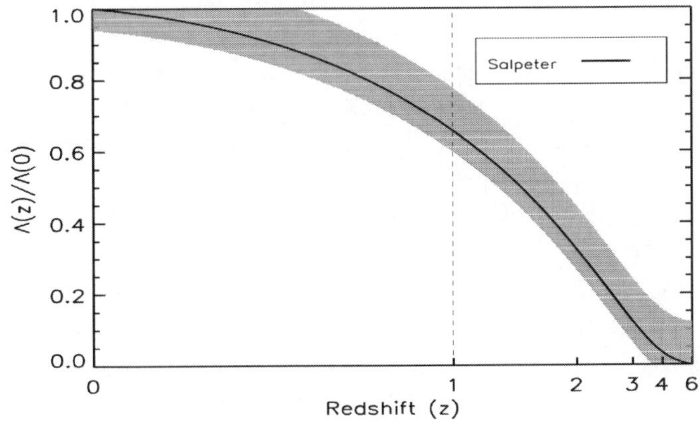


Figure 9.1. The normalized redshift evolution of the dark energy density, denoted by $\Lambda(z)$, normalized to its present-day value $\Lambda(0)$. The shading represents the observational uncertainty in deriving the black hole formation rate, when a universal Salpeter initial mass function is not strictly valid.

## 1.3. The Spatial Scale and Timing of Dark Energy

The current value of the dark energy density, $\Lambda(0)$, is measured to be $\sim 10^{-29}$ g cm$^{-3}$, according to the Planck satellite again. Because $\Lambda = 2N_{\rm BH}m_P/L_f^3$, the relation between the dark energy density and black hole number can be used to compute $L_f$, as follows. The present number of macroscopic (mostly stellar origin) black holes in the observable universe is roughly $N_{\rm BH} \sim 10^{19}$.[2]

[1]Of course, the attentive reader may think of inflation driven by dark energy. Indeed, mini wormholes can do this too, as is discussed further below.

[2]Recall that there are about $10^{11}$ galaxies with each about $10^{11}$ stars. Roughly one in $10^3$ stars ends its life as a black hole.

From the present-day value of the dark energy density it then follows that $L_f \approx 2 \times 10^{14}$ cm, about the size of our solar system. Hence, the physical scale of the dark energy density, not specified in the topological dynamics approach studied here, is determined by empirical data.

A number of other popular concerns regarding dark energy find a natural resolution when macroscopic black holes induce mini wormholes. Because the star formation history of the universe sets the appearance of present-day dark energy, it is ultimately general relativity itself that determines the evolution of $\Lambda$ (or some effective $w$). This renders the occurrence of the cosmological "constant" in Einstein gravity much less ad hoc, and even appropriate. Furthermore, since black holes are stellar remnants, it should come as no surprise that one witnesses dark energy in our time. Today is part of an epoch during which the bulk of the stars in the universe has been formed. This while the most massive of those stars ($> 8M_\odot$) have already exploded as type II supernovae to leave behind black holes. i.e., the change in the magnitude of the current dark energy density is set by the number of macroscopic black holes. With the bulk of all stars formed, so with much of the baryonic matter in galaxies turned into stars, this number cannot increase much further. Mankind thus finds itself, while orbiting a typical star, close to the maximum dark energy density in time.

This is for some value of $L_f$. The dark energy density is roughly comparable to the current total matter density. This may seem to be an odd coincidence. However, a larger characteristic volume $L_f^3$ for the causally connected universe when black hole $X$ forms, diminishes $\Lambda$ as much as it increases the total mass in the observable universe available to make black holes after that time. It is quite reasonable that for a larger mass in a cosmos the same processes describe the formation of stars. So the resulting number of black holes, relevant to $\Lambda$, is then mostly affected by the mean mass of a black hole combined with the fraction of baryonic matter that is converted into stars and subsequently into black holes. For the present universe these numbers are about $\mu = 3M_\odot$ and $s = 10^{-3}$, respectively. If they are universal, then $\Lambda$ is not affected.

Stellar physics forbids much lower mass black holes because electron or neutron degeneracy pressure provides stability (Pauli's exclusion principle). A much larger black hole mass $\mu$ as a function of cosmic time (so redshift) would require a very exotic stellar initial mass function, which is not observed. This while the probability with which massive stars yield black holes is not likely to vary much around $s$ either. In fact, one can equate $\Lambda(0) = 10^{-29}$ g cm$^{-3}$ to $2m_P N_{\rm BH}/L_f^3 = 2m_P sY(\rho_c L_f^3)/(\mu L_f^3)$, for the mass in the observable universe written as $\rho_c L_f^3$, which yields $N_{\rm BH}$ black holes through $s$, $\mu$ and the star formation probability $Y$. The explicit dependence on $L_f^3$ now drops out and is replaced by the characteristic mass density $\rho_c$. This assumes that the bulk of the mass in the universe has had the opportunity to enjoy a four-space history that allows it to undergo a star-forming process with probability $Y$, and one that may reach a density $\rho_c$ at some time. In practice this means that gas must have accumulated and formed galaxies over billions of years.

The density $\rho_c$ is roughly $3 \times 10^{12}/Y$ g cm$^{-3}$, and indicates what local environment is responsible for black hole formation, so for dark energy generation. In this, $Y \sim 0.01 - 0.1$ for the typical efficiency with which gas is converted into stars, according to observations. Interestingly, $\rho_c$ matches the density for which a high-mass star enjoys a collapse of its core and can subsequently evolve only into an object supported by neutron degeneracy pressure or into a black hole. Indeed, $\rho_c$ is precisely the stellar evolution quantity that should generate dark energy as the result of black hole formation. Therefore, the link between dark energy and stellar origin black holes renders the former sensitive to cosmic evolution in a manner that is dictated by quite basic and well understood astrophysics: the numbers $\rho_c, s, Y, \mu$. Hence, the value of $\Lambda(0) \sim 10^{-29}$ g cm$^{-3}$ is robust and universal, within an order of magnitude, for an epoch after the peak in the cosmic star formation rate.

Conversely, the mean cosmic density with its $(1+z)^3$ dependence (any linear scale goes as $1+z$) is currently about a few $\times 10^{-30}$ g cm$^{-3}$ and increases by an order of magnitude into the past, e.g., to $z \sim 2$. This while the universe has spend most of its cosmic life after the peak in the star formation history. This peak occurs for $z = 1 - 2$, at least nine billion years ago, but also billions of years after the big bang. The densities of dark energy and matter are roughly comparable for $z \sim 1$, when the bulk of the stars/black holes are in place. Also, the Sun with its planets is likely to find itself after the peak in the cosmic star formation rate, on statistical grounds, if the Sun is just one of many stars formed in the past nine billion years.[3] The latter favors, through cosmic expansion, the dark energy density to be higher by only a modest factor relative to the total

[3]The Sun is about five billion years old, relatively recent but very reasonable in a probabilistic sense.

matter density today.

Hence, both the timing and magnitude of $\Lambda$ are as one would expect. An anthropic principle does not appear to be needed in any strong sense. The anthropic principle states that observers, like human beings, detect the universe the way it is because precisely those properties allow for their existence. Had the Earth been formed already a few billion years after the big bang, before the peak in the cosmic star formation history, then $N_{\rm BH}$ had simply not yet reached its maximum. Dark energy would not be completely absent either to those putative observers, simply a bit weaker. In any case, our presence in the current epoch is statistically likely. So no fine-tuning is needed.

## 2. The Wormhole Nature of Inflation

The inflationary idea has shaped modern-day cosmology and knows many different incarnations. In it, the universe undergoes an early phase of very rapid expansion that leads to its current state of isotropy/homogeneity and flatness in terms of spacetime curvature. To further constrain topological dynamics, this section aims to link the solutions of the equation of motion above to a possible inflationary phase of the early universe.

### 2.1. The Number of e-Foldings

In the early universe, before black hole $X$ provides a sense of longevity for the quantum foam, equation (8.17) presents an *exponential* solution of the equation of motion for the wormhole/handle multiplicity. Just like Wheeler's induced quantum foam can be associated with a slowly evolving dark energy density in recent times, so can the young and rapidly growing quantum foam phase prior to the occurrence of black hole $X$ be used to drive inflation. After all, as pairs of mini wormholes are created each Planck time their number is driven to about one per Planckian volume $l_P^3$. Higher values cannot be attained because merging becomes important then and this violates the assumption of $\Delta M = 0$ underlying equation (8.17).

This means that the dark energy density carried by induced mini wormholes prior to black hole $X$ is approximately $\Lambda_P \sim m_P/l_P^3$, with $P$ referring to Planck. Clearly, the negative gravity of spacetime itself overwhelms all mass density contributions then. Consequently, the four-space vacuum drives the expansion of our cosmos so strongly that the size of the universe increases exponentially with time. This is called a de Sitter universe. Indeed, to any observer Planck size wormholes are induced and embedded every Planck time and in every Planck volume. So one expects that a local measure of size $R$ of the universe obeys $R/(dR/dt) \sim t_P$ under those conditions, with an obvious exponential solution. Conversely, the present-day dark energy density is only a factor of a few larger than the total mass density, with the latter about 120 orders of magnitude below $\Lambda_P$ (see further down). So the current accelerated expansion of the universe is also exponential, but on a much longer time scale of billions of years because of the sparser density of induced mini wormholes.

It is nice that one can solve two problems with the same physical object: the black hole. Of course, for inflation to commence and continue, so for equation (8.17) to pertain, it is necessary that the ambient matter and radiation density are high enough to suppress the evaporation of quantum black holes. This is because the accretion of mass and energy lowers the black hole temperature to below the Planck energy (so $\Delta E = 0$ for equation (8.17) then). Furthermore, as mentioned above, close to the Planck density merging between quantum black holes occurs, and this also increases their mass. The latter inevitably increases black hole longevity, which automatically leads to the appearance of a black hole $X$ that lives for longer than the contemporary age of the universe and thus the freeze-in of $L_f$. As such, there is a natural end to wormhole driven inflation through the scale change $l_P \to L_f$, which thus constitutes a graceful exit. Indeed, the ratio $l_P^3/L_f^3 \sim 10^{-140}$. For $\sim 10^{19}$ black holes in the observable universe this explains why $\Lambda_P$ and $\Lambda(0)$ can be 120 orders of magnitude different.

Through topological dynamics it is possible to compute from $L_f$ the number of e-foldings n, so the factor $e^{\rm n}$ by which the universe grows during inflation, as follows. First realize that the topologically void sub-equation of motion for $S^3$ is actually useful geometrically. Recall that

$$\Delta n_2 S^3 = \delta_2 n_2 S^3 = 0, \tag{9.4}$$

and consider topological time in the limit $m >> 1$ and $\Delta m >> 1$, written as $\tau$ for $m$ and $d\tau$ for $\Delta m \leq m$, during inflation. So the value of $\Delta m$ need not be small when using the variable $\tau$. In physical units $\tau = mt_P >> t_P$, with an associated physical scale $R(\tau) = c\tau = cmt_P = ml_P$. One can now turn to the *geometrization* of equation (9.4). Although yielding a zero in terms of counting loops, geometrically one can represent the three-sphere by the relative size of the universe $S(\tau)$ for $n_2(m) \rightarrow S(\tau)$. That is, $S(\tau)$ measures the ratio (a dimensionless number) of the size of the universe with respect to its pre-inflationary state. Any notion of geometry on $S^3$ should allow one to do this if time is intrinsically discrete and if the speed of light $c$ in vacuum is a universal constant. i.e., one simply tallies line segments in units of $l_P$.

The action of $\Delta$ is a first order time difference, which can be written as $dS/d\tau$, for changes that are not necessarily small. The action of the counting operator $\delta_2$ only has meaning in conjunction with $n_2(m) \rightarrow S(\tau)$ and must yield $\delta_2 S \rightarrow 0$ when one takes the limit $S \rightarrow \infty$. Indeed, inflation is a homeomorphism that drives the universe to a flat state and inflation must end when $S$ is large. So the geometrization of $\delta_2$ represents (diminishing) curvature expressed through $S$. Consequently, $\delta_2 \rightarrow 1/S^2$ because this is the curvature of a surface $S^2$ on which an observer resides during expansion. One finds that equation (9.4) turns into

$$\frac{dS}{d\tau} = \frac{1}{S}. \tag{9.5}$$

The solution $S(\tau) = (2\tau_f)^{1/2}$ is easily found when one integrates from 0 to the final time $\tau_f$ that marks the end of inflation because black hole $X$ is formed. Since $\tau_f = L_f/l_P$, one obtains $S_f = 2^{1/2}(L_f/l_P)^{1/2}$. Note that $L_f \approx 1.3 \times 10^{47}$ in units of $l_P$, so that $S_f \approx 5.0 \times 10^{23}$. That is a very large increase in the size of our universe, worthy of being called inflation indeed. For $S_f = e^{\mathrm{n}}$ one finds $\mathrm{n} \approx 55$. This value is comfortably within the fiducial range measured by the already mentioned Planck satellite, $50 - 60$.

## 2.2. The Scaling of Cosmological Perturbations

Induced mini wormholes appear in a quantum spacetime $L(T^3)^+$ that enjoys an exponentially rapid increase in physical scale $R(\tau)$. This picture introduces temporal effects and can be used to determine the so-called tensor-to-scalar ratio r and the spectral index $n_s$, both numbers that cosmologists like to determine very much. The number **r** measures the relative strength of gravitational waves, so metric perturbations on a $0 - 1$ scale, that are produced during inflation. The number $n_s$ describes how the power of fluctuations is distributed over length scale $l$ through $l^{-(n_s-1)}$. Hence, $n_s = 1$ is a scale-free distribution. Note that both these numbers, like n above, are dimensionless. Hence, topological dynamics can say something about them as well through n.

The two quantities **r** and $n_s$ respectively express the occurrence of gravitational waves and the deviation from scale-free behavior in mass density fluctuations. The parameter **r** derives from the fact that induced wormholes come and go. Hence, they constitute fluctuations in local mass density that create ditto fluctuations in curvature, which in turn travel as a causal signal at the speed of light. The parameter $n_s$ represents the fact that exponential expansion comes with a characteristic scale, basically n. In this, topological time $\tau$ and geometric time $t$ are not the same to an observer for $\Delta m \approx m >> 1$. This is because of the topological identifications (portals through time) that act as reality inductors and allow the global exchange of observers and observed. So there is a fundamental distinction between local (geometric time) and global (topological time) observers. Without going into much mathematical detail, one can determine **r** and $n_s$.

Consider a local observer during inflation who is comoving with the exponential expansion of the universe. Then $t = ln(R(\tau)) = ln(\tau)$ in natural units, with large topological time steps $\Delta m$ taken globally. Consequently, the local observer notes that induced mini wormholes are spread out through time because the universe expands over n e-folds. i.e., Wheeler's quantum foam is not well localized in time due to the exponential stretching of the lattice of three-tori. The observer thus sees mini wormholes appear and disappear, generating a gravitational wave signal as they distort quantum spacetime, with a duty cycle of $1/\text{n}$. i.e., set by the total duration of inflation in his/her frame of reference.

Furthermore, every three-torus is mimicked six-fold each Planck time, *with* geometric perturbations and all. A global observer detects this as a seven times stronger perturbation in his/her time frame. However, the strength of this multiplication effect equates to $ln(7)$ as a local observer detects and mimics it because $t = ln(\tau)$. Hence, combined this gives $\mathbf{r} = ln(7)/\text{n} \approx 0.036$ for the gravitational wave signal as we see it today. According to recent Planck results, $\mathbf{r} < 0.11$ and one finds comfortable consistency. Folding in other observational limits, e.g., from the BICEP2 experiment on the south pole[4], yields a very ten-

[4]This telescope covers a much smaller part of the sky than the Planck satellite. So one should be careful with Milky Way interstellar dust that emits polarized radiation of the same signature

tative determination of $\mathbf{r} \approx 0.05 \pm 0.03$. It will be interesting to watch these observational numbers improve in the coming years.

Because inflation is exponential and n is much larger than unity, there is hardly any length scale dependence of **r**, i.e., only a logarithmically weak one. Still, according to the MP, the characteristic (so-called "pivot") scale $l_f$ on which these perturbations cross our Hubble radius should obey $l_f/L_f = 2^{1/2}S_f^{1/2}$ and $l_f \approx 10^{26}$ cm. That is, the frozen-in scale $L_f$ (about the size of the solar system) that determines $S_f$ is itself mimicked through equation (9.5) into a scale $l_f >> L_f$ (about the size of a supercluster of galaxies) that sets the largest gravitational structures in our cosmos. Indeed, mimicry pertains to scaling laws as well.

To determine the relation between **r** and $n_s$ one simply realizes that **r** is equal to the relative amplitude of curvature perturbations associated with mini wormhole induction on a dynamical $L(T^3)$ background. This is *equal to* the deviation from scale-free behavior that any observer measures in that $L(T^3)$ background if the induced mini wormholes are used as a frame of reference spanning n e-folds. After all, $L(T^3)$ is scale free and its topologically and geometrically perturbed state is $L(T^3)^+$. Subsequently, **r** and $n_s$ are related by $n_s = 1 - \mathbf{r} \approx 0.964$. The latter value agrees very well with recent Planck observations, which indicate $n_s = 0.9603 \pm 0.0073$. Also, topological dynamics has no free parameters to change the values of **r** and $n_s$. Indeed, n is set by the empirical number $N_{\rm BH} \sim 10^{19}$ (the total number of black holes in the observable universe) through the dark energy density $\Lambda = 2N_{\rm BH}m_P/L_f^3 \approx 10^{-29}$ g cm$^{-3}$ as measured today, which thus fixes $L_f$ in the expression for $S_f = e^{\rm n} = 2^{1/2}(L_f/l_P)^{1/2}$.

Finally, mass density fluctuations are generated by a large number of evaporating mini wormholes in topological dynamics. It is therefore likely that the law of large numbers renders the statistics of the resulting probability distribution close to Gaussian. Also, such induced mini wormholes are mimics that appear throughout causally connected spacetime. In this, any such four-space region must always be perceived as the mimic of some other, conveniently linked to each other through an $S^1$ loop. Of course, a full exploration of equation (8.12), also using general relativity, is required to confirm these assertions for the early universe and to reproduce the properties of the cosmic microwave background.

---

as cosmic microwave photons perturbed by gravitational waves.

Nevertheless, it is easy to see that the relative amplitude of the mass density fluctuations described above must be $\sigma \sim (2N_{\rm BHII})^{-1/2}$. This for a population of $N_{\rm BHII}$ longer-lived black holes that maintains Wheeler's induced quantum foam at $\Lambda_P$, but does not contain black hole $X$ yet, as follows. Given the discrete nature of induced mini wormhole pairs, the underlying statistics is shot noise. Hence, for $N_{\rm BHII}$ transient wormhole pairs, $\sigma$ is as above. Furthermore, grand unified theory has a characteristic energy scale of $E_c \sim 10^{16}$ GeV, or about $10^{-3}$ of the Planck energy. During this epoch, the universe has substantially cooled down in energy relative to $m_P c^2$, while primordial black hole production may naturally occur during the associated $m_P c^2 \to E_c$ phase transition. The temperature $T_{\rm BH}$ of a black hole scales with its mass $M$ as $T_{\rm BH} \propto 1/M$, and thus its typical life time $t_e$ as $t_e \propto M^3$. It is therefore reasonable to estimate on thermodynamic grounds that one has longer-lived mini black holes of mass $m_{\rm BHII} \sim 10^3 m_P$ that live for about $t_e \sim 10^9 t_P$.

Also, $m_{\rm BHII} \sim 10^3 m_P$ yields $l_t \sim 10^3 l_P$ for the typical transient size scale of dark energy prior to freeze-in of $L_f$. During the epoch of primordial longer-lived black holes, $\Lambda(m) = 2N_{\rm BHII} m_P / l_t^3$ is kept at the Planck energy density. This implies that $N_{\rm BHII} \sim 10^9$ for $l_t \sim 10^3 l_P$. The latter number of $10^9$ should be close to the maximum value and thus set $\sigma$. An energy density of roughly $\Lambda_P$ strongly suppresses the evaporation of induced mini black holes, thereby increasing the number of longer-lived BHs $N_{\rm BHII}$. However, the accretion rate of already existing longer-lived black holes would also be very high at the Planck energy density. The latter boosts the mass of these black holes, and thus their longevity since $t_e \propto M^3$. Hence, $N_{\rm BHII} \sim 10^9$ is not likely to grow much before a black hole $X$ appears with a life span[5] that exceeds the age of the universe at that time, thus freezing in the value of $L_f$. In all, one finds a value of $\sigma \sim 2 \times 10^{-5}$ for the magnitude of cosmological mass-energy fluctuations, in agreement with current observations. Given that $N_{\rm BHII}(m) >> 1$, the Poisson distribution of this shot noise is indeed typically indistinguishable from Gaussian noise.

[5]Black hole $X$ may not have survived until the present day. Nevertheless, its existence is a part of our universe's past forever.

# 3. Quantum Spacetime Contributions to Black Hole Growth

One may also use the MP to explore the quantum properties of black holes further. In this, it is good to recall that the MP provides a simple rule to apply: an observer may mimic the effects undergone by an observed. To this effect, one chooses a globally distinct yet locally equivalent path through quantum spacetime. With the construction of $L(T^3)^+$ this is now straightforward, really. First though, it is prudent to discuss some more black hole astronomy.

## 3.1. Astronomical Context

Almost every galaxy is known to harbor a supermassive black hole in its nucleus, with masses ranging from $10^5$ to more than $10^{10}$ times the mass of our Sun, indicated again with the symbol $M_\odot$. Best known is the so-called Magorrian relationship. It states that the mass of the central black hole is $\sim 10^{-3}$ of the stellar bulge mass, albeit with a large scatter. The bulge refers to the central concentration of stars that is typically present in elliptical and spiral galaxies (with names that simply refer to their morphology in starlight). Furthermore, black holes with masses as large as $10^9\ M_\odot$ have been detected around a redshift of $z \sim 6$. This leaves about a billion years since the big bang to grow such monsters. In any case, the growth of black holes (be they large or small) is of fundamental astrophysical importance.

In recent years, many efforts have been devoted to understand the growth of black holes, both today and in the early universe. Traditional black hole growth proceeds through an accretion disk around the black hole that feeds matter to the event horizon by viscous effects, as described earlier. One may also explore the possibility that quantum spacetime fluctuations, in the form of induced mini wormholes, allow the mass growth of black holes on cosmologically interesting time scales. Such an unconventional growth mode operates in the absence of any ambient matter and allows macroscopic black holes to feed off the mini wormholes they induce.

## 3.2. Autonomous Growth of Quantum Black Holes

The macroscopic length scale characteristic of topological dynamics is $L_f$, the size of the universe when the first black hole forms that lives longer than the contemporary age of the universe. This black hole was called $X$ above. It must have a typical mass $m_X$ and thus a lifetime $t_X$ under Hawking evaporation. This time scale $t_X$ is a quite relevant global quantity, like $L_f$ is. It sets the time interval $m' = t_X/t_p = \Delta m$ over which the first global topological identification through time is made. i.e., the one that establishes the actual future identity of our universe through an observer's mimicry of black hole longevity. This notion of the global future being linked to a local observer may seem bizar, but it should be no surprise by now. For a black hole to have any existence, it needs to be entangled with an observer. When this situation pertains to time scales that are relevant to the universe itself, then also the future identity of the whole must derive from the mimicry of such a present part.

In this, the typical mass of black hole $X$ is likely to be a strong fluctuation with respect to the mean mass of the black hole collective during inflation. This is because it is the creation of black hole $X$ that ends inflation. Furthermore, besides the already encountered Planck energy of $m_P c^2$, particle physics suggests that a characteristic energy during inflation is $E_c \sim 10^{-3} m_P c^2$. This is basically the grand unification energy scale for which the electromagnetic, weak and strong force are unified, but gravity has already split off. As discussed before. Recall that it is thus plausible that particle-like entities such as black holes have roughly a mass of $10^3 m_P$ on thermodynamic grounds because then their $T_{\text{BH}} \approx E_c$. In all, for an excursion from the $E_c$ mean by a factor of 10, this yields $m_X \sim 0.3$ g and therefore an evaporation time $t_X \sim 4 \times 10^{12} t_P$ for black hole $X$. This is only about $10^{-31}$ s of course...

Still, the Machian scales $t_X >> t_P$ and $L_f >> l_P$ are global and *partition* quantum spacetime whenever observers interact with quantum black holes, as an expression of non-local inertia. Note that $t_X$ is much larger than $t_P$, but very small compared to human and even sub-atomic experience. This while $L_f$ represents how any path in a time slice of thickness $t_X$ through four-space is spatially sub-divided, with $L_f$ about the size of our solar system and equal to $A = L_f/l_P \approx 2 \times 10^{47}$ Planck lengths. The crucial point is that our universe has expanded to a size many times larger than $L_f$, while the current age of the cosmos is much larger than $t_X$.

The above implies that the probability to find an induced mini wormhole along a path segment of length $l_P$ through a leaky black hole is $A^{-1} \approx 5 \times 10^{-48}$. This is in a time slice of thickness $t_X$. Consequently, the time scale $t_Q = At_X \approx 10^9$ years quantifies how long it takes to actually encounter the induced mini wormhole inside the quantum black hole that induced it. Of course, this argument is heuristic. However, one may also use the MP to relate quantum spacetime fluctuations in the form of mini wormholes to observers that study black hole event horizons.

To this effect, imagine a distant observer who measures the mass $M$ of a black hole, similar to the example in chapter 5 involving black hole temperature. The observer mimics an induced mini wormhole of mass $m_P$ if he/she chooses a four-space round-trip from $L(T^3)^+$ that takes an $S^1$ loop passing *through* a black hole horizon, yielding an actual wormhole topology. The observer's path is then globally distinct from a route that traverses the mini wormhole itself because there are two topologically distinct leaky black hole/mini wormhole paths (one with mass $M$ and one with mass $m_P$). Yet, within measurement uncertainty, the observer's path is also locally equivalent to a path that does pass through the mini wormhole.

Being in the Newtonian regime again, the mimicking observer simply detects the mass of the macroscopic black hole, and not that of the induced mini wormhole, with respect to some distant flat region of three-dimensional space. However, this detected mass cannot be a static quantity anymore since the mimicked mini wormhole is highly transient. Indeed, it is a quantum mechanical fact that the mini wormhole must appear, and thus locally add its mass $m_P$ to the inducing black hole as a pertinent topological entity, during its lifetime only. Of course, this life span is rather short (about a Planck time), contrary to the evaporation time of the macroscopic black hole used by the observer for his/her mimicry.

Hence, according to the MP there is no way around it: the addition of the mini wormhole's mass carries over to our macroscopic world and the observer must measure an increase in the mass of the black hole used for his/her mimicry that is equal to the original mass $M$ of this black hole. That sounds disturbing and rather mind boggling. Nevertheless, it is an unavoidable consequence of the simple fact that macroscopic black holes mimic mini wormholes as much as they induce them. Quite right, the collectives of macroscopic black holes and induced mini wormholes are entangled, and thus the former may feed off the latter.

Fortunately, the pertinent global information in this case, and the salvation of a more or less well behaved universe, is the time required to accumulate such a quantum mimicry mass. Indeed, $t_Q$ is a comfortable billion years and sets how long the observer needs for his/her mimicry. i.e., how long it takes on average to encounter an induced mini wormhole, mimicked as a macroscopic black hole of mass $M$, along the round-trip above. So $t_Q$ sets how rapidly perturbations in the Schwarzschild radii $R_s$ of quantum black holes grow, independent of their masses.

Consequently, the act of black hole observation chooses a four-space history characterized by a mimicry enhanced black hole mass. One may wonder whether a human observer on Earth physically goes on a billion years trip through a black hole to see it grow. In a sense, the answer is yes. Any observer, from a particle to a human being, interacts with the universe through its very existence and thereby chooses a global history via the topological structure of quantum spacetime. Put differently, any observer of a macroscopic black hole also performs his/her actions as a mimicker of a mini wormhole, whether he/she likes it or not. So in a fleeting quantum sense, all observers do make such round-trips.

In any case, every observer is bound to the global partitioning of the quantum foam and needs to take $A$ steps of duration $t_X$ to complete the temporal $S^1$ loop through a black hole in $L(T^3)^+$. Therefore, one has a change in horizon scale

$$\delta R_S = R_S/t_Q\ \delta t, \tag{9.6}$$

for some coordinate time interval $\delta t$. Because $R_S \propto M$, for black hole mass $M$ in solar masses, one finds the infinitesimal mass evolution

$$\delta M/\delta t = M/t_Q. \tag{9.7}$$

This yields a quantum accretion rate $R_Q$ of

$$R_Q(t) = M(t)/t_Q \quad M_\odot\ \mathrm{year}^{-1}. \tag{9.8}$$

i.e., for times longer than $t_Q$, black holes significantly feed off their quantum foam and gain mass. Thus, black holes spontaneously lower their thermodynamic temperature $T_{\mathrm{BH}} \propto 1/M$ and become colder. The fact that the interaction strength $t_P/t_Q \sim 10^{-60}$ is tiny, highlights the intrinsic weakness of this *autonomous* growth. So when viewed as an instability, it is of very modest importance. However, our universe has survived for so long (about 14 billion years) that Wheeler's induced quantum foam of mini wormholes is pertinent in a cumulative sense.

Interestingly, $t_Q \approx 10^9$ years is only a factor of ten larger than the time scale of $t_E \sim 0.45\epsilon/(1-\epsilon)$ billion years for so-called Eddington-limited accretion, with $\epsilon \approx 0.1$ the radiative efficiency encountered in chapter 6. The latter time scale is a theoretical estimate for how long it takes black holes to roughly double their mass by accretion if the power output from their accretion disk is only slightly less than that needed to push back the matter flowing to the event horizon. The numerical similarity between $t_Q$ and $t_E$, and the fact that black holes are not likely to sustain high accretion rates for long, renders autonomous growth a cosmologically pertinent mechanism. Autonomous growth alone, so without traditional mass accretion, follows

$$M(t \geq t_1) = M_1 e^{(t-t_1)/t_Q}, \tag{9.9}$$

with $t_1$ the time at which a black hole is formed and $M_1$ its mass at that time. The current age of the universe is about 14 billion years ($\sim 10^{61} t_P$). Hence, the boost factor $\mathbf{B} = e^{t/t_Q}$ can be as large as $\sim 10^6$, thus allowing for very significant autonomous growth. A fiducial time scale of 10 billion years, so $z \sim 1.5$ at the peak of the cosmic star formation history, still yields $\mathbf{B} \approx 2 \times 10^4$. This all seems very relevant for our universe and worthwhile to explore further.

### 3.3. Astronomical Consequences of Autonomous Black Hole Growth

One expects stellar origin black holes of $1 - 30\ M_\odot$ and monolithic ones of $10^4 - 10^6\ M_\odot$ for a universe of a few hundred million years old, as discussed in chapter 6. Therefore, a monolithic black hole of $10^5\ M_\odot$ at redshift $z = 20$ (roughly 140 million years after the big bang) quantum grows by a factor of $\approx 8 \times 10^5$ during the age of the universe. It thus reaches the scale of the biggest supermassive black holes known today ($> 10^{10}\ M_\odot$). A stellar origin black hole of 3 $M_\odot$ can grow from $z = 10$ (roughly 400 million years after the big bang) to a current mass of $2 \times 10^6\ M_\odot$. Also, the mass of the black hole in our own Milky Way center is about $4 \times 10^6\ M_\odot$. Hence, half of that mass can be acquired through autonomous growth of a stellar mass black hole. Quite tantalizing, the Milky Way black hole should quantum grow by $\sim 4 \times 10^{-2}\ M_\odot$ in 10 years due to induced mini wormholes. This growth is slow, but significant over cosmic times. In fact, sensitive telescopes (so-called interferometers) should be able to detect autonomous growth for black holes that do not feed much off a more conventional accretion disk.

Supermassive black holes at high redshift benefit least from autonomous growth. This is because the characteristic time scale $t_Q$ allows for only modest growth (a factor of 2-3) by $z \sim 6$ since the big bang. This strengthens the need for massive monolithic black holes of $10^{4-6}\ M_\odot$ at redshifts beyond 10 and/or very efficient accretion onto stellar origin black holes, aided by galaxy mergers under the influence of gravity.

On the other hand, the present-day Magorrian relationship is helped along by these black hole quantum effects. Autonomous growth attaches a minimal gain to the initial mass function of black holes, irrespective of the efficiency of mass accretion through a disk or the mergers of galaxies. This while star formation is a global process that, because it involves Jeans unstable gas clouds at the bottom of a galaxy's potential well, is likely to sustain a steady mean rate over times longer than $t_Q$. This way a bulge of stars is formed naturally. Of course, black hole driven outflows may even trigger star formation through shocks that compress gas clouds and thus trigger star formation. Under these circumstances, autonomous growth alone may not allow the black hole mass to keep up with the stellar bulge mass, but such star formation episodes are also much shorter than $t_Q$. Overall, autonomous growth at least supports co-eval evolution of a central supermassive black hole and the stellar population of its host galaxy, with some scatter. Furthermore, autonomous growth allows supermassive black holes ($> 10^{10}\ M_\odot$) to exist in otherwise normal galaxies, so violating the Magorrian relation in that the central black hole is overly massive. Such monsters have actually been observed.

A very important final consequence of autonomous black hole growth is the following. One finds that $\mathbf{B} > 10^{12}$ some 14 billion years from now, so when the age of the universe has roughly doubled. At this future time, autonomous growth renders central black holes more massive than their host galaxies, even if all supermassive black holes today have a stellar origin. Therefore, quantum grown black holes eventually overcome all the other mass in the universe... They would even start to dominate dark energy, since the latter depends on black hole number alone and remains constant with time if no additional black holes are formed. Indeed, most galaxies have converted the bulk of their mass into stars by now, so that the very long-term future of the universe is one with constant $\Lambda$ (negative gravity), but ever increasing gravitational pull by quantum grown black holes. Consequently, a so-called *big crunch* becomes the fate of the universe, rather than eternal expansion. One should realize that this is due to the specific $t_Q$ value pertinent to the four-space history of our universe. That fate is one of many, but it is the one that we are subject to.

## 4. Local Mass and Global Mass: the Higgs Boson

To close off this chapter, it is good to turn to the masses of elementary particles. Recall first that mimicry causes three-tori in $L(T^3)^+$ to form septuplets. This fact was already used in the discussion on mini wormhole inflation to determine r and $n_s$. Indeed, the impact of mass carrying mini wormholes in the quantum spacetime vacuum is linked to the multiplicity of thee-tori. Hence, whatever interactions occur on $L(T^3)^+$, as the topological origin of inertia, their relative strength when it comes to local entities like particles is only $\frac{1}{7}$ of that in a simply connected spacetime. This is because there are always seven three-tori for a particle to occupy.

Independent of this, particle physics has its own way of attributing mass to particles. This is done by an elementary particle called the Higgs boson. Interactions with the Higgs boson give mass to all other elementary particles. Since mini wormholes behave like non-local particles themselves, one can imagine that elementary particles traveling through $L(T^3)^+$ pick up wormhole interactions and thus inertia. Of course, this is a collective process, where all wormholes in $L(T^3)^+$ contribute. One then better speaks of a field that assigns mass to particles, the so-called Higgs field, as they traverse it.

Furthermore, wormholes also evolve and affect each other as discrete entities. After all, they can evaporate and merge, thereby changing the nature of the collective. In particle physics one then says that the Higgs field interacts with itself. The individual excitations of the collective (so of the field) are particles and called Higgs bosons. In topological dynamics these are mini wormholes. Obviously, a wormhole $S^1 \times S^2$ comprises a single loop and its particle identifier is therefore that of a boson (see chapter 4). Indeed, Higgs bosons are the local particle physics description of the global mini wormhole ensemble. In this, the availability of Higgs bosons for material interactions is only limited by the total number of mini wormholes induced over the entire past history $m \leq m_0$ of the universe, so comprising about $10^{61} t_P$.[6]

Interestingly, one can express the mass of this Higgs boson as $m_{\rm H} = (2v^2\lambda)^{1/2}$, with the experimentally determined Fermi coupling constant $G_{\rm F} = 1.16632 \times 10^{-5}$ GeV$^{-2}$ $= v^{-2}/2^{1/2}$.[7] So the value of $v$ in the expression for $m_{\rm H}$ is known. This does not hold for the so-called coupling constant $\lambda$, at least within particle physics. However, the value of $\lambda$, as its name suggests, expresses with what efficacy the Higgs boson can provide mass to particles in the universe. This is a global process. Hence, it must be a direct consequence of how effectively mini wormholes funnel the mass that they carry into the lattice of three-tori. Consequently, one has that $\lambda = \frac{1}{7}$. i.e., $\lambda$ is set by the number of degrees of freedom of an individual three-torus in a septuplet and mini wormhole mass is simply watered down.

This value $\lambda = \frac{1}{7}$ yields a Higgs boson mass $m_{\rm H} = 131.6$ GeV. It should be no surprise that $m_{\rm H}$ is so much smaller than $m_P$. After all, the topological identifications on $L(T^3)^+$ spread the mass of induced mini wormholes out. Recent Large Hadron Collider results from CERN yield an experimental number of $m_{\rm H} \approx 125$ GeV, different by about 5%. Not bad, but one should do better. Indeed, there are subtleties in what can be called an observer mass or an observed mass for the Higgs boson. After all, the Higgs boson gives mass to *itself* as well, calling for the observer-observed entanglement that lies at the heart of this monograph. Indeed, the distinct global four-space paths that wormholes as a collective provide also define their individual local masses on scales much larger than $l_P$. That notion should be scrutinized, as follows.

The computation above does not make explicit use of the MP when it comes to assigning mass. Recall first that Mach's principle formed one of the great motivations for Einstein's general theory of relativity: the inertia of matter locally is determined by the mass distribution of the universe globally. However, in order to quantify particle masses generated by a wormhole collective, in some local inertial frame, one needs to be able to measure relative to a region that is not

[6]This contrary to the dark energy density, which derives from $N_{\rm BH}(m)$ and $L_f^3$.

[7]Recall that one GeV is approximately equivalent to $1.6 \times 10^{-24}$ g times $c^2$.

affected by that process. A Higgs boson is both a representation of a mini wormhole and an observable entity. This implies that its mass cannot be detected in a strictly local sense. After all, a mass measurement, as are all observations, is intrinsically non-local according to the MP. i.e., any observer needs to mimic the mass setting experience of a Higgs boson through a globally distinct yet locally equivalent four-space path. The magnitude of the required non-locality can be quantified through a dimensional argument. The latter is an order of magnitude estimate guided by the correct physical units of an effect, and goes as follows.

One should first note that the Higgs boson is associated with the electroweak force. This force is broken into its present-day electromagnetic and weak part below an energy of $E_{\rm H} = 246$ GeV. This so-called symmetry breaking leads to the global occurrence of the Higgs boson as a particle, so as an *individual* entity that carries mass rather than just the wormhole collective (so the field). For isotropic spacetime, all observers living now can receive information on a portion of the universe with a size that is equal to the particle horizon, $R$, equal to the distance over which matter can travel during the evolution of our cosmos (currently billions of light years). With Planck's constant $h$ and speed of light $c$, the ratio $D = RE_{\rm H}/(hc) \sim 10^{44}$ then constitutes the maximum dimensionless extent, so the number of path segments, over which any observer can speak of the Higgs boson. Indeed, measurements on scales smaller than $l = hc/E_{\rm H}$ cause restoration of the electroweak symmetry because $l$ corresponds to an energy $E_{\rm H}$. Consequently, the value of $D$ cannot be exceeded during Higgs boson measurements. Therefore, $D = RE_{\rm H}/(hc) = L'/(cT')$ should generally hold for the scale $L'$ up to which the Higgs boson is detectable locally by observers for a duration $T'$. Even for $T' = t_P$, the value of $L'$ is truly macroscopic and about $10^6$ km. This appears perfectly comfortable because $L' >> l_P$ clearly holds and $L'$ is also much larger than any sub-atomic scale.

However, the path that an observer needs to take, to obtain information on the Higgs mass, is that of a topological identification through time. e.g., by the use of an $S^1$ loop in $L(T^3)^+$ to transport the Higgs boson to a mass measuring apparatus and to transport it back. This route is topologically distinct, because it loops through four-space, but locally the same as the path through the CERN accelerator. Unavoidably, this path of the mimicking observer leads to a Higgs particle outside of the accelerator, with some probability $p$, because $L'$ is so large. Indeed, within a four-dimensional spherical region $U$, centered on the experiment and of linear size $\leq L'$, the probability density to enjoy a topological identification through time between the center of $U$ and another spatial point scales like $1/r^3$. This is for an observer with a measuring apparatus residing at a distance $r$ from the center of $U$ and on its spatial boundary. The probability to find a Higgs boson outside of a radius $r$, with $cT' < r < L'$ and integrating over shells $4\pi r^2 dr$, is then $p(r) = log(L'/r)/log(L'/(cT')) = log(DcT'/r)/log(D)$. Conveniently, this is only weakly dependent on details that affect $L'$ or $r$ due to the logarithmic dependence.

For $L' \sim 10^6$ km, so if $T'$ is as small as the Planck time (appropriate for our universe), one finds a probability $p(r) \approx 18\%$ to encounter Higgs particles beyond $r \sim 10$ m from an experimental event. This is quite significant, given that a detector region of $r \approx 10$ m seems reasonable for modern-day facilities (like the CMS and ATLAS detectors inside the Large Hadron Collider). In any case, the dependence of $p$ on $r$ is logarithmic and this *leakage* by topological identifications is therefore a robust consequence of topological dynamics. There is one more step to take now.

Given that the Higgs boson assigns mass to itself, the topological identifications on $L(T^3)^+$ must affect the Higgs boson mass as determined by a distant observer as much as its spatial appearance in terms of $r$. That is, the mimicking route of the observer spreads out the spatial presence of the Higgs boson and therefore also the mass that it may endow itself with...

Realize then that this topological leakage is a four-space phenomenon. Any mass scale $M_0$ has an associated four-space volume $V_0 = 1/M_0^4$ (so in natural units). This $V_0$ is rescaled to $V = (1-p)^{-1}V_0$ through the topological identifications on $L(T^3)^+$. i.e., one of the directions of a local four-dimensional cube is stretched out by a factor $1/(1-p)$. One therefore has a globally observable Higgs mass $M_0 = 131.6$ GeV, so for $p = 0$ as $r \to L$. While the locally observable Higgs mass that results from mimicking is $M = V^{-1/4} = (1-p)^{1/4}M_0 = 125.2$ GeV for $p(r) = 18\%$ and $r = 10$ m. Because one deals with statistics this local mass $M$ of the Higgs boson has an associated probability of $1 - p \approx 82\%$ to be detected. With a more modest probability of 18%, so $(1-p)/p \approx 4.6$ times less likely, an observer can still detect the global mass $M_0$.

Obviously, in the limit of large $r$, when the scale of the detectors inside the measuring apparatus approaches $L'$, the value of $p(r)$ goes to zero. Only the global mass is then measured because all spread out Higgs boson events (so also the decays of Higgs bosons) can be captured. In any case, the Large Hadron Collider value of $\approx 125$ GeV corresponds very well with $M$. It would of course be interesting to vary the size $r$ of the detector and correlate it with the measured Higgs boson mass. Also, a $4.6$ times weaker secondary signal at a Higgs mass $M_0$ should occur in the experiments at CERN.

# Chapter 10

# Future Outlook: The Multiverse

Time to finish this monograph and to end with a few final thoughts. The central object in this monograph is the path and its uses through the MP. The crucial role played by three-tori and wormholes, with long-lived leaky black holes inducing transient mini wormholes, then leads to the name of topological dynamics for the resulting theory. The quantum spacetime $L(T^3)^+$ yields many interesting results as direct applications of the MP: a global four-space history understanding of quantum entanglement (the double slit experiment and Schrödinger's cat), an alternative derivation of black hole temperature, dark energy induced by quantum black holes, the nature of wormhole driven inflation and fluctuations in mass density, coarse grained discrete time, autonomous black hole growth, and the mass of the Higgs boson. Perhaps the simplest yet most revealing formula of topological dynamics is equation (8.14), $\Delta E = 2n_1$; the number of transient mini wormholes $\Delta E$ equals twice the number of macroscopic black holes $n_1$.

Most importantly as far as the underlying philosophy of topological dynamics is concerned, it is found that topological identifications in the form of $S^1$ loops act as reality inductors because they allow the exchange of observers and observed. These portals through time on $L(T^3)^+$ therefore play a crucial role in the way that Nature expresses quantum processes. Indeed, the essence of topological dynamics is that Nature distinguishes all phenomena through geometric and topological information, where the latter is highly non-local and dynamical. Hence, the very nature of $L(T^3)^+$ suggests, as has been suspected by many, that quantum mechanics is even more mysterious than its original inception almost a century ago would suggest. In fact, since the topological identifications on $L(T^3)^+$ are dynamical, and are already required in an essential manner for known microscopic and macroscopic processes, one should determine what other consequences quantum spacetime might have. This raises the question of where to go next.

It is good to return then to the essence of the MP. An observer and observed are distinct yet equivalent because one can play the role of the other, while both are necessary to define reality. That is, an observer can freely mimic the experience of an observed, and vice versa, while observers and observed must collectively carry reality. In this, Nature is fundamentally undetermined, and this measurement uncertainty affects spacetime dynamics. Consequently, connectivity and not shape characterizes quantum four-space, as well as its interacting observers and observed.

Hence, a multiply connected spacetime emerges in which topologically distinct paths are both the consequence of observations and the cause for observed experiences. In it, any observer can mimic any observed. In fact, mimicry is the only action that constitutes an actual observation, and thereby defines reality, in quantum spacetime. Mimicry therefore effectuates a physical existence by inducing a history that is locally equivalent to, and globally distinct from, an observed's four-space path. The sought after distinction in this comes about by using multiply connected spacetime to choose a path that involves a portal through time (so an $S^1$ loop that constitutes a reality inductor). Such a new four-space history is a simple exchange of the observer and the observed in terms of experienced local events. Yet it is also beyond the reach of a continuous spacetime deformation of the original history in a global sense, thereby adding a new fiber to the universe's four-space fabric.

The moral of the MP story is then the following: everyone can be an observer or an observed and only relative merit can be assigned to either distinction. Consequently, the collective existence of observers and observed derives from one mimicking the other unremittingly. In such a world of identity by mimicry, any local piece of history is globally acquired by the induction of some new part of four-space. So in a very real sense this implies that time itself is multiply connected, in both a microscopic and macroscopic manner. The popular notion of a multiverse, as many parallel yet independent universes that are much like the many worlds of Everett, then takes a fundamentally new form.

One in which all the topologically distinct histories in four-space are globally entangled and fully interactive at the discretion of a mimicking observer! Pertinent examples to this effect have been discussed in previous chapters, like the quantum growth of macroscopic black holes and the non-local mass of the Higgs boson.

The global induction that is part of the MP does not violate causality. After all, the latter holds locally just the same. Rather, time is already multi-fingered through general relativity. Quantum effects then further render it a global patchwork of time lines, with many interconnections that carry observer-observed entanglement through the topological dynamics of four-space. One can go further still. An observer who mimics, as the only way of performing an observation, relocates from one quantum history to another by inducing the latter and by using *only* information. This is not time travel in the traditional sense. It signifies that a single act of observation modifies the self-induced structure $L(T^3)^+$, and thus causes a global change in the reality called our universe. This merely by mimicking the quantum entanglement information needed to describe reality. Information that is available to all observers.

Indeed, given the common big bang origin of our universe, each entity in it is entangled with all others when it comes to their four-space histories. Such is the nature of unremitting mimicry. One may object that earlier examples on the role of information in gravity showed that sometimes only a part of the local particle information $N$ and the global entanglement information $N'$ matters for the strength of gravitational interactions (depending on circumstances). This while the induction of an entire distinct yet equivalent history is required, with a mimicry identifier $N'_O = 2$, if an observer and observed need to be defined individually for the specific interaction and system under consideration. Certainly, for an everyday world of distant galaxies, observed and mimicked by an observer on a planet Earth ($N'_O = 2$) that also orbits the Sun ($N'_O = 1$), the entanglement between observer and observed is not immediately apparent, and quantum effects due to the MP can be large or small.

However, the $N'_O = 1$ case encountered in this monograph is general relativity in its local geometric formulation, without global quantum information. Clearly, the MP is intrinsically self-inducing and the entangled existence of observers and observed as distinct yet equivalent entities is simply always present at the quantum level. After all, it is through Einstein's elevator thought experiment that a mimicking rocket operator emerged. Hence, even though certain physical processes can be described without the induction of entire histories, so through classical physics and with high accuracy, there is no escaping the MP as a general quantum rule. Inevitably, there is just no objective reality. Nature mimics invariably through the induction of locally equivalent yet globally distinct histories.

Furthermore, these considerations also hold for human beings and the neurons that carry their consciousness. For the mind's eye of an observer toward the universe, it is obvious then that a distinct yet equivalent history must always be induced as well, if one is to speak of an observer and the observed universe. The key point here is that the MP only requires local equivalence and allows for any global distinction, without a priori limits on the latter's magnitude in terms of physical phenomena. In fact, the MP effects in this monograph run quantitatively from, e.g., a few percent for the Higgs boson mass, to a factor of six for the information corrected gravitational constant $G' \approx 6G$, to effectively unbound masses for autonomous black hole growth.

Consequently, one should then also entertain the following speculative thoughts, from a philosophical perspective. Instead of performing traditional quantum mechanical actions, an observer can equivalently interact only with his/her own perception of reality. He/she thereby alters material existence in a tangible manner by inducing a globally distinct yet locally equivalent four-space path. Indeed, in the MP the nature of the observer and the observed is arbitrary. Such an induced history of mental origin need not even be a small change with respect to the mimicked one in terms of the energy required to achieve it by more traditional means. After all, only the addition of quantum entanglement information matters, and induction could therefore yield a significant reality reset.

This can be accomplished, but it would be necessary to wrap one's mind around $L(T^3)^+$ in a literal four-dimensional sense. In life, it is not essential what one does, but only how one does it. The same holds for quantum spacetime, as the structure that dictates how physical processes occur, with their quantitative expression subject to fundamental measurement uncertainty. Clearly, quantum spacetime must also include the ponderings of any conscious observer in this, given that each of us is a particular collection of $S^1$ loops. The lattice of three-tori with wormholes is then the fundamental expression of both self-awareness[1] and tangible reality. This renders the global properties of $L(T^3)^+$ receptive to local cognitive processes if one mimics its *path* connectivity. This is an interesting idea to pursue indeed..., but not for this monograph. In closing, I wish the reader all the best.

[1]A self-contained process of selection and distinction, so of choice.

# Suggestions for Further Reading

This monograph aims to present results derived from first principles. Hence, no references are explicitly cited in the main text. However, the interested reader may find the books and papers mentioned below of use when exploring various topics further.

A few classical textbooks with self-explanatory titles:

Misner, C.W., Thorne, K.S., & Wheeler, J.A., 1973, "Gravitation"; [Freeman]
Nakahara, M., 1990, "Geometry, Topology and Physics"; [Hilger]
Zeilik, M. & Gregory, S.A., 1998, "Introductory Astronomy & Astrophysics"; [Thomson]

A few classical papers by famous physicists:

Einstein, A., 1915, Preuss. Akad. Wiss. Berlin, Sitzber., 778 (general relativity)
Feynman, R.P., 1948, Rev. Mod. Phys., 20, 367 (path integral formulation)
Wheeler, J.A., 1957, Annals of Physics, 2, 604 (quantum foam)

Some background work by the author:

Spaans, M., 1997, Nuc. Phys. B, 492, 526 (multiply connected spacetime)
Spaans, M., 2013, Int. J. Mod. Phys. D, Vol. 22, 1330022 (topological dynamics)

# Index